Surface Mount Technology

Other McGraw-Hill Reference Books of Interest

Handbooks

Avalone and Baumeister • STANDARD HANDBOOK FOR MECHANICAL ENGINEERS
Beeman • INDUSTRIAL POWER SYSTEMS HANDBOOK
Coombs • BASIC ELECTRONIC INSTRUMENT HANDBOOK
Coombs • PRINTED CIRCUITS HANDBOOK
Croft and Summers • AMERICAN ELECTRICIANS' HANDBOOK
Di Giacomo • VLSI HANDBOOK
Fink and Beaty • STANDARD HANDBOOK FOR ELECTRICAL ENGINEERS
Fink and Christiansen • ELECTRONICS ENGINEERS' HANDBOOK
Harper • HANDBOOK OF THICK FILM HYBRID MICROELECTRONICS
Hicks • STANDARD HANDBOOK OF ENGINEERING CALCULATIONS
Inglis • ELECTRONIC COMMUNICATIONS HANDBOOK
Johnson and Jasik • ANTENNA ENGINEERING HANDBOOK
Juran • QUALITY CONTROL HANDBOOK
Kaufman and Seidman • HANDBOOK FOR ELECTRONICS ENGINEERING
  TECHNICIANS
Kaufman and Seidman • HANDBOOK OF ELECTRONICS CALCULATIONS
Kurtz • HANDBOOK OF ENGINEERING ECONOMICS
Stout • HANDBOOK OF MICROPROCESSOR DESIGN AND APPLICATIONS
Stout and Kaufman • HANDBOOK OF MICROCIRCUIT DESIGN AND APPLICATION
Stout and Kaufman • HANDBOOK OF OPERATIONAL AMPLIFIER DESIGN
Tuma • ENGINEERING MATHEMATICS HANDBOOK
Williams • DESIGNER'S HANDBOOK OF INTEGRATED CIRCUITS
Williams and Taylor • ELECTRONIC FILTER DESIGN HANDBOOK

Electronics Technology and Packaging

Edosomwan • IMPROVING PRODUCTIVITY AND QUALITY FOR ELECTRONICS
  ASSEMBLY
Seraphim, Lasky, and Li • PRINCIPLES OF ELECTRONIC PACKAGING
Coombs • PRINTED CIRCUITS HANDBOOK
Woodson • HUMAN FACTORS REFERENCE GUIDE FOR ELECTRONICS AND
  COMPUTER PROFESSIONALS

Dictionaries

DICTIONARY OF COMPUTERS
DICTIONARY OF ELECTRICAL AND ELECTRONIC ENGINEERING
DICTIONARY OF ENGINEERING
DICTIONARY OF SCIENTIFIC AND TECHNICAL TERMS
Markus • ELECTRONICS DICTIONARY

For more information about other McGraw-Hill materials,
call 1-800-2-MCGRAW in the United States. In other
countries, call your nearest McGraw-Hill office.

Surface Mount Technology

Materials, Processes, and Equipment

Carmen Capillo
UNISYS Corporation
Network Computing Group
San Jose, Calif.

McGraw-Hill Publishing Company

New York St. Louis San Francisco Auckland Bogotá
Caracas Hamburg Lisbon London Madrid Mexico
Milan Montreal New Delhi Oklahoma City
Paris San Juan São Paulo Singapore
Sydney Tokyo Toronto

Library of Congress Cataloging-in-Publication Data

Capillo, Carmen
  Surface mount technology: materials, processes, and
equipment / Carmen Capillo.

  p. cm.
  Includes index.
  ISBN 0-07-009781-X
  1. Printed circuits—Design and construction.   2. Surface
mount technology.   I. Title.
  TK7868.P7C36   1990
  621.381'531—dc20

89-12699
CIP

1234567890   DOC/DOC   9876543210

ISBN 0-07-009781-X

The editors for this book were Daniel A. Gonneau and Jim Halston,
the designer was Naomi Auerbach, and the production supervisor was
Richard A. Ausburn. This book was set in Century Schoolbook by the
McGraw-Hill Publishing Company Professional & Reference Division
composition unit.

Printed and bound by R. R. Donnelley and Sons

*For more information about other McGraw-Hill materials,
call 1-800-2-MCGRAW in the United States. In other
countries, call your nearest McGraw-Hill office.*

To my parents and, wife, Mercedes

Contents

Preface

Since the 1960's, Surface Mount Technology (SMT) has been considered an absolute necessity by military and aerospace industries in order to achieve the highest electronic densities and performances. Today all types of commercial industries from automobiles to home appliances consider the use of SMT not only necessary from a product standpoint but also from a competitive one.

This electronic packaging technology, which consists of many integral technologies, not only can provide the user with very high packaging densities and circuit speeds but also an enormous amount of challenges. The technology behind these challenges is what I have written about in this book. I have dispensed with generalities and instead have provided hands-on technical details related to design, materials, processes, manufacturing techniques, equipment, workmanship, and test within SMT based on years of multimillion-dollar R&D programs and many more years setting up and maintaining various types of surface mount assembly/end product factories.

This book consists of five parts. Part 1 covers the range of Surface Mount Devices (SMDs) from chip components to burn-in sockets and packaging. Part 2 provides design-related information such as land (pad) geometries, thermal management, and how design affects manufacturability, reliability, and costs. Part 3 provides specifics in manufacturing flows and processes as they relate to the type of surface mount designs. In addition, the technology and various test methods behind incoming inspection of manufacturing materials (adhesives, solder pastes, fluxes, etc.), PCBs, and SMDs are provided in great detail along with numerous graphic acceptance guidelines. Part 4 provides two separate chapters covering soldering and cleaning. Each chapter provides backbone information on the various reflow and cleaning processes along with their respective process techniques, machinery, materials, process controls, and problem/solution related information. Testing surface mount assemblies is discussed on an electrical engineering level in Part 5. Again, generalities are missing in this chapter, which provides information relating to assembly faults, fixturing, in-circuit, and functional testing.

CARMEN CAPILLO
UNISYS Corporation
Network Computing Group

Acknowledgments

I wish to acknowledge several colleagues for their help and support in the preparation of this book; Bill Deforest, Hughes Aircraft, for valuable discussions regarding PCB fabrication; George Capillo, Hughes Aircraft, for his technical review and assessment of the page proofs; Ken Jessen, Hewlett-Packard Company, for extensive technical information on test; Steve Robeson for the difficult photographic reproductions, and special thanks to Al Coe, UNISYS/Convergent Technologies, for his continuous support during the preparation of this book.

I wish to strongly emphasize that no endorsement of any machines and materials is implied through their presentation in this book. Many thanks are extended to those who took the time to provide me with illustrations and photographs.

Acknowledgments

Introduction to Surface Mount Technology

1.1 The History of Surface Mount Technology

The early stages of surface mount technology (SMT) occurred in military and aerospace electronics during the mid-1960s. The surface mount device (SMD) which made this possible is called a *flat pack*. The flat pack, as shown in Fig. 1.1, can be considered the father of SMDs. It consists of a ceramic package with flat ribbon leads, called *gull-wing* leads by today's terminology. This SMD simply replaced the dual-in-line package (DIP) and was and still is used for high-density memory cards commonly found in navigation and communication systems aboard military aircraft. At this time, the major reason that flat packs were used instead of DIPs was the ability to place these components on both sides of the printed circuit board (PCB), thereby practically doubling the electrical performance. This was, and still is, the major reason why the military and aerospace industries utilized SMT: it produces substantially higher circuitry densities (performance) per

Figure 1.1 A ribbon-leaded flat pack mounted in a plastic carrier. The standard military and aerospace leaded surface mount IC package.

1

square unit area per assembly than the conventional through-hole technology can provide.

In the early 1970s another SMD became available called the *hermetic leadless chip carrier* (HCC or LCC). This was the next effort to provide even higher levels of circuitry density for military and aerospace applications. This component, however, presented reliability challenges in that the leadless approach removed the required compliancy (or stress relief) necessary when ceramic, inorganic components are mounted directly onto typical organic PCBs. While extensive research was being performed to resolve this compliancy problem, a temporary surface mount approach utilizing LCCs was being utilized. This consisted of mounting LCCs onto a ceramic substrate, thereby matching the coefficient of thermal expansion (CTE) between that of the LCC and the substrate, which negates the compliancy problem. This technique was considered temporary and was called the *daughter board* approach. As shown in Fig. 1.2, the daughter board consisted of a small ceramic substrate with LCCs mounted onto both sides. This assembly was then mounted onto the "mother board" in the same way as DIPs. However, some military and aerospace companies, such as Hughes Aircraft, Martin Marrietta, and Tracor Aerospace, could not wait for the compliancy problem to be resolved and were forced to attempt to use large ceramic substrates with LCCs mounted onto them. As shown in Fig. 1.3, this type of assembly made the advantages of SMT extremely visible; when LCCs were utilized, multiple conventional assemblies could be designed in one smaller surface mount assembly (SMA).

The large ceramic substrate had its fabrication limitations, which

Figure 1.2 A daughter board with a ceramic substrate and LCCs mounted onto both sides.

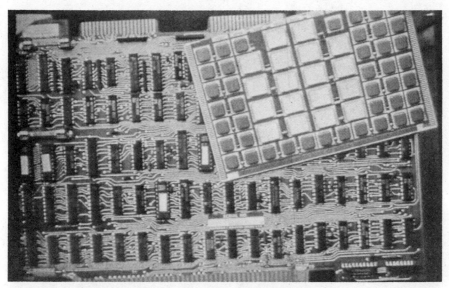

Figure 1.3 In the background, a conventional through-hole-mounted assembly. In the foreground, a surface mount assembly with four times the electrical performance as the conventional assembly.

kept the driving forces behind the effort to resolve the compliancy problem between LCCs and organic PCBs. By 1976, breakthroughs in the technology of producing organic PCBs with tailorable and controllable CTEs resulted in major redesigns and new designs of a wide range of military and aerospace assemblies incorporating the LCC. The standard assembly, as shown in Fig. 1.4, consisted of epoxy Kevlar or polyimide–Copper-Invar-Copper (CIC) PCBs with LCCs.

By the late 1970s, commercial companies started to announce the availability of various plastic surface mount packages, such as the SOT 23, SOIC, and PLCC, which could use the standard PCB technology and offered the obvious advantages of higher densities. From that point, the automobile and computer industries started to design new products and redesign old products utilizing these new plastic SMDs. These two large industries started the real takeoff and the fever that still exists today relating to the use of SMT and its inevitable superiority over conventional through-hole technology.

1.2 An Explanation of SMT

Electronic assemblies which utilize SMT are those that are populated with surface mount devices (SMDs). SMDs, sometimes called surface mount components (SMCs), can be defined as any electrical component, leaded or leadless, mounted on one or both sides of a substrate

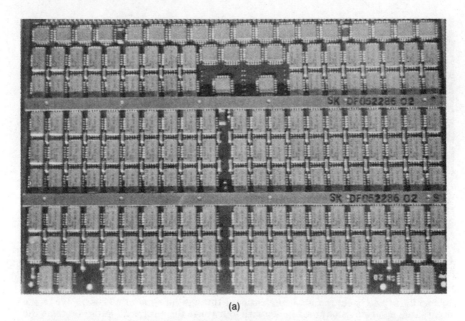

(a)

(b)

Figure 1.4 (a) A typical military and aerospace SMA with LCCs mounted onto a polyimide-CIC multilayer PCB; (b) close-up photograph of the same SMA with chip capacitors located between the LCCs demonstrates the high packaging densities capable with SMT.

and interconnected to the substrate lands or metallizations via a lap solder joint. In contrast, a conventional component is connected to the substrate via a through-hole solder joint where the component's leads make metallurgical connections on both sides of the substrate and within the plated through-hole.

A SMD makes metallurgical contact only with the substrate side onto which it is mounted. This is what makes SMT an attractive assembly technique; unlike conventional assemblies or components, both sides of the substrate can be populated with components. Furthermore, with as much as 50 to 70 percent size reductions, SMDs are significantly smaller than their equivalent conventional types.

1.3 Applications of SMT

The phenomenal growth of SMT has led to its usage in nearly all types of electronic products in most of the industries existing today. Some of the high tech applications consist of satellite and manned spacecraft as shown in Fig. 1.5. Figure 1.6 illustrates SMAs of various designs commonly used in these high tech applications. The commercial industry, which is now one of the largest users of SMT, initially incorporated SMT into some of the first portable business computers released to the marketplace in 1982–1983. Figure 1.7 illustrates a very compact portable business computer having both conventional components and SMDs. Today, the computer industry utilizes SMAs for high-speed circuitry, memory boards, liquid crystal display assemblies, disk drive assemblies, and many other parts of the computer and its accessories. Some of the latest applications in computers where SMAs have reached an impressive stage can be seen in Fig. 1.8 where high-pin-count SMDs are used for VLSI technology.

Medical electronics is yet another commercial industry that is becoming more and more dependent on the advantages of SMT and at the same time maintaining the necessary levels of reliability. In this industry, SMT is finding its way in medical products such as intravenous feeders, glucose measuring equipment, pacemakers, and numerous other types of electronics. Figure 1.9 illustrates a SMA which makes up the main electronics in one of the most advanced pacemakers. This assembly incorporates a variety of SMDs such as a crystal, a socket, a transformer, and SOICs.

1.4 Current Advantages and Disadvantages of SMT

Contrary to what some people believe, SMT is not a new, unproven technology. As mentioned, it has been used in military and aerospace

(a)

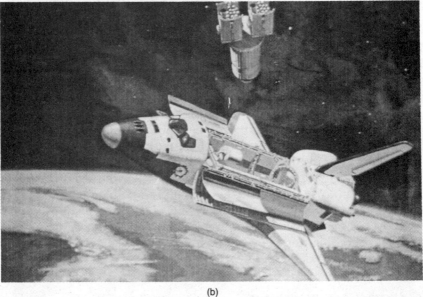

(b)

Figure 1.5 (*a*) A Hughes Aircraft Company communications satellite; (*b*) Rockwell International's space shuttle.

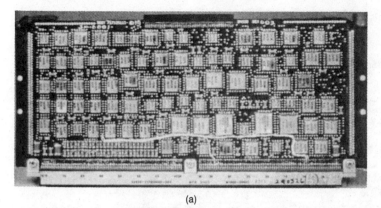

(a)

(b)

Figure 1.6 SMAs of various designs used in military and aerospace applications. (a) A lightweight, high-heat-dissipating SMA; (b) a so-called mother board SMA with LCCs of various sizes.

electronics for the past 20 years, primarily because of its higher packaging density which also benefits in less total weight and faster electrical speeds. Because of SMT's extensive use in high-technology and life support applications, hundreds of millions of dollars have been spent via government grants and private industry funding to study and thoroughly understand all of the variables which affect its reliability and manufacturability. Therefore, making the decision as to whether or not SMT is reliable and manufacturable is no longer nec-

Figure 1.7   One of the first portable business computers which extensively utilized conventional technology and SMT on the same PCB. (*Courtesy Convergent Technologies.*)

Figure 1.8   The latest advances of SMT used in high-speed supercomputers incorporating leaded SMDs with 340 leads per component on 0.020-in. centers.

essary, especially from a commercial usage standpoint. This is simply because billions of dollars worth of hardware, some of it in orbits 23,000 miles above the planet and some of it going beyond our solar system, conclusively demonstrate the reliability and manufacturability capable of SMT. Certainly satellites costing in excess of

Figure 1.9   A SMA used in the pacemaker industry.

$300 million each and space probes over $1 billion each would only incorporate SMT if it was a viable technology.

Commercial usage of SMT is currently going through its adolescent phase, with both advantages and disadvantages as compared to conventional component through-hole mounting. The advantages can be given as follows:

1. Fifty to seventy percent higher packaging densities.
2. The possibility of consolidating several conventional assemblies in one SMA.
3. Higher pin counts possible on components.
4. Higher performance speeds possible due to significantly shorter lead lengths.
5. No component preparation (lead forming) required prior to assembly.
6. More and faster automated manufacturing.
7. Less storage space for components required.
8. Less manufacturing space required.
9. Lower overall costs.

The disadvantages can be given as follows:

1. Lack of standards for components, design, processes, materials, workmanship, and testing.

2. Lack of significant numbers of professionals and suppliers with experience in SMT.

3. Higher initial costs for most design, components, and assembly.

4. Absence of some components in surface mount packages.

5. Significantly more complex manufacturing assembly processes and materials.

6. In-circuit test more complicated pending design efficiency.

Upon review of the list of advantages and disadvantages of using SMT, one will start to realize that as the technology matures in the commercial industry, its disadvantages are basically only temporary conditions and are constantly being worked on.

1.5   A Decision Model for Choosing SMT

Choosing whether or not to implement SMT in new products or even to redesign old products should not be as difficult a decision as most people tend to make it. If three basic conditions exist, one should be led into SMT:

1. If the electronic assembly requires higher packaging density so that it can all fit a fixed or limited space.

2. If performance such as high speed, high frequency, reduced noise, and reduced weight that is required on a product can only be achieved by using SMT.

3. If competition is producing products (either now or soon) at quality, cost, and performance levels that cannot be matched using conventional technology.

Ultimately, within a certain period of time, SMT, like all leading technologies, will become the standard manufacturing technique, leaving behind the old ones. Furthermore, as greater packaging densities are required, technologies such as TAB and chip-on-board may start becoming more dominating technologies than SMT.

Surface Mount Components

Introduction

The lap solder joint and the through-hole solder joint are what separate the technology of surface-mounted and conventionally mounted components. To explain further, for each lead or metallization attached to the component, SMDs have solder joints only on the "surface" of the substrate material. Conventional components, which are always leaded, have a topside, a through hole, and a bottom side solder joint for each lead. An exception to these conditions exists where the leads on conventional components can be formed so that they are surface-mountable. Leads on dual-in-line packages (DIPs) can be formed into an inward or outward lap configuration and butt joint configuration. Axial and/or radial leaded components can be typically formed outward into a surface mount configuration.

  Although these techniques are performed by cautious newcomers or those who have component availability problems, it is not a new trend since some important factors cannot be overcome by surface-mounting conventional components. These important factors include the reduction in size, material, and printed circuit board (PCB) real estate for attachment. As later compared in detail, conventional components such as DIPs, electrolytic capacitors, resistors, and inductors are normally three to five times larger in size when compared to equivalent types of SMDs. Also, because of the through-hole solder joint configuration, conventional components require space on both sides of the substrate material; SMDs do not.

  Since the 1960s, SMDs have originated and played an

important role in military and aerospace technologies, which required high-reliability electronic packaging and assemblies. When high density became a relevant factor, these high technology assemblies were typically populated with ribbon-leaded ceramic flat packs instead of the larger through-hole-mounted ceramic DIP packages.

The main reasons for these high technology industries to capitalize on SMT in its earliest stages were to obtain higher packaging and functional densities. If these two purposes were not incorporated, space, weight, and reliability limitations would make advanced machines and electronics such as fighter aircrafts, satellites, and space vehicles usually impossible or impractical. Costs, in these cases, have typically been a secondary factor.

Upon the introduction of a SMD called the ceramic leadless chip carrier (LCC) into the marketplace during the early 1970s by the 3M Company, the military and aerospace companies immediately saw great advantages to these new components over the widely used ceramic flat-pack components for packaging integrated circuits. Because of this, government- and company-sponsored research and development programs created a competitive race for understanding the materials, processes, and reliability of these LCC assemblies. At the same time, the costs of these new components became a more concerning issue since high technology electronics were already at astronomical prices. Promotional efforts and additional competition by other manufacturers who started to produce LCCs established efforts to produce these components at lower costs and in commercial quantities and styles. Also, other factors that promoted SMDs starting with the LCC, were the SMT seminars, papers, and rumors. Industrial rumors like "soon LCCs will become less expensive than flat packs and easier to handle" and "soon you will only be able to buy LCCs, whereas flat packs and DIPs will become premium cost components due to lead forming, preparation, and handling" were not uncommon.

Furthermore, to make the introduction of SMT and LCCs more favorable, equipment manufacturers began to produce faster, more accurate component placement and mass reflow soldering systems for LCCs and other types of SMDs that started to emerge. Possibilities of faster production rates and lower costs caught the attention of consumer and industrial product manufacturers, which resulted in manufacturers packaging all types of SMDs in leaded and leadless surface-mountable plastic and ceramic styles.

Passive and Discrete Surface Mount Components

2.1 Passive Components

2.1.1 Electrical resistance

To some degree, all materials have a certain amount of opposition to a flow of current. This property in materials is called *electrical resistance* or simply resistance. Resistance is measured in ohms and given a symbol named for the German scientist, George S. Ohm (1787 to 1854), who studied the phenomenon or relationship that explains the voltage, current, and resistance of an electrical circuit. This relationship can be expressed by the following equation, known as Ohm's law:

$$V = IR$$

Here R is the resistance of the material connected across the terminals of the voltage source, V is the voltage, and I is the current. By knowing the value of any two of these quantities, the third can be derived from the law.

There are great differences in the resistances of various materials of the same size and shape; for example, carbon has a resistance about 2000 times greater than copper, and glass about 100 billion times higher than that of carbon. The current that actually passes through materials such as glass is so small that these materials are considered insulators of electricity. The resistance of the material depends not only on its composition but also on its shape. The smaller the area through which current flows, the greater the resistance. As we would also expect, the resistance increases both with the length of the current path and as the temperature of the material increases.

2.1.2 Chip resistors

The resistor is the simplest type of electronic component, and its electrical property is to restrict or oppose the flow of current. A surface-mounted chip resistor consists of a leadless rectangular aluminum oxide (Al_2O_3) chip with two external terminations called *end caps*. Compared to conventional resistors, which are normally comprised of cylindrical carbon composites or films with protruding axial leads on both ends, surface-mounted chip resistors of equivalent value can be 70 percent smaller in size. Although size is an obvious advantage with chip resistors, unit cost and assembly flexibility can add a significant reduction of cost when compared to conventional parts.

There are two different types of chip resistors: commercial and military. Two specifications currently govern the requirements for chip resistors: EIA IS-30 is for commercial usage and MIL-R-55342 is for military and aerospace usage. Both specifications provide dimensional, electrical, and physical performance requirements along with specific test methods and procedures.

The EIA IS-30 has three standardized sizes, the RCO805, RC1206, and RC1210 types, although one can obtain approximately 11 different sizes of resistors ranging from the extremely compact 0302 series to the very large 2512 series. Figure 2.1 lists the physical dimensions for the three EIA standardized types of commercial chip resistors. For high-volume production and standardization, the author recommends the 1206 series, which lend themselves to easy packaging and machine placement and handling. The small types, such as the 0805, typically cause more difficult manufacturing concerns such as feeding, pickup, and adhesive bonding problems; whereas the larger types, such as the 1210, typically are not always readily available and are more easily damaged due to their large surface area.

Chip resistors in conformance to MIL-R-55432 exist in five different sizes, ranging from the smallest RM0505 size, which measures 0.505×0.050 in., to the largest RM2208 size, which measures

Style	Body length	Body width	Body thickness	Termination width	Rated wattage	Maximum continuous working voltage
	(L)	(W)	(T)	(t)	(mW)	(V)
RC0805	0.079 +/− 0.008 (2.00 +/− 0.20)	0.049 +/− 0.008 (1.25 +/− 0.20)	0.020 +/− 0.008 (0.50 +/− 0.20)	0.016 +/− 0.010 (0.40 +/− 0.25)	62.5	100
RC1206	0.126 +/− 0.008 (3.20 +/− 0.20)	0.063 + 0.006/−0.008 (1.60 +0.15/− 0.20)	0.020 +/− 0.010 (0.56 +/− 0.15)	0.022 +/− 0.006 (0.50 +/− 0.25)	125	200
RC1210	0.126 +/− 0.008 (3.20 +/− 0.20)	0.098 +/− 0.008 (2.50 +/− 0.20)	0.22 +/− 0.006 (0.56 +/− 0.15)	0.20 +/− 0.010 (0.50 +/− 0.25)	250	200

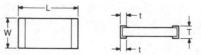

Figure 2.1   The three standardized chip resistors in conformance with EIA interim standard IS-30. Dimensions in inches (millimeters).

Dimensions

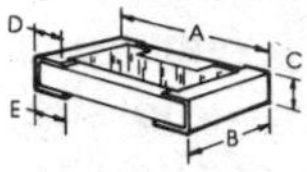

Size	RM0505		RM0705		RM1005		RM1505		RM2208	
Length A	0.050	+0.025 / -0.005	0.075	+0.025 / -0.005	0.100	+0.025 / -0.005	0.150	+0.025 / -0.005	0.225	+0.025 / -0.005
Width B	0.050	+0.010 / -0.005	0.050	+0.010 / -0.005	0.050	+0.010 / -0.005	0.050	+0.010 / -0.005	0.075	+0.010 / -0.005
Thickness C	0.010/0.040		0.010/0.040		0.010/0.040		0.010/0.040		0.010/0.040	
Top end band D	0.010	±0.005	0.015	±0.005	0.015	±0.005	0.015	±0.005	0.015	±0.005
Bottom end band E	0.015	±0.005	0.015	±0.005	0.015	±0.005	0.015	±0.005	0.015	±0.005

Resistance ranges

Size	RM0505	RM0705	RM1005	RM1505	RM2208
Max. power voltage	50 mW, 40 V	100 mW, 40 V	100 mW, 40 V	150 mW, 40 V	225 mW, 40 V
Tolerance	1% 2% 5% 10%	1% 2% 5% 10%	1% 2% 5% 10%	1% 2% 5% 10%	1% 2% 5% 10%

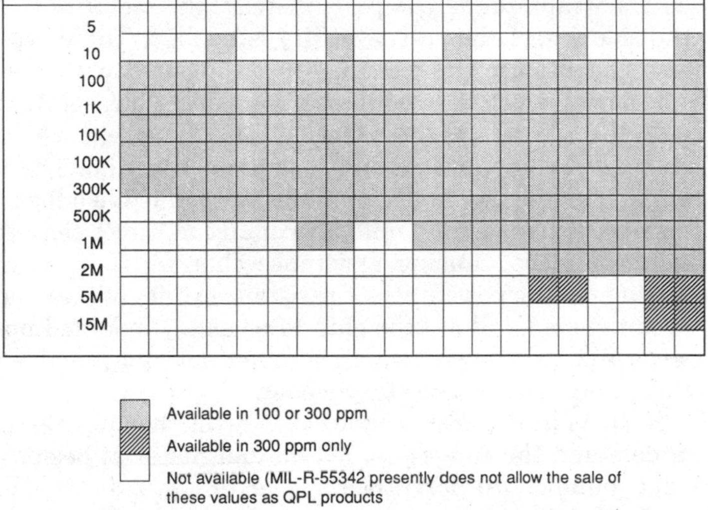

(Resistance value chart with rows: 5, 10, 100, 1K, 10K, 100K, 300K, 500K, 1M, 2M, 5M, 15M)

▒ Available in 100 or 300 ppm

▨ Available in 300 ppm only

☐ Not available (MIL-R-55342 presently does not allow the sale of these values as QPL products

Figure 2.2   The five standardized chip resistors in conformance with MIL-R-55432.

0.225 × 0.075 in. Figure 2.2 lists the physical dimensions and resistance ranges for all five sizes of military chip resistors. It should be mentioned that due to its square shape, type RM0505 does not allow mechanical determination of the termination or end cap location. Therefore, this part is difficult to automatically measure, orientate, and package.

2.1.2.1   Chip resistor construction and materials.   The surface mount resistor is available in two forms, the *chip* and *metal electrode face bonding (MELF)* packages. The chip type, discussed in this section, is the most widely available and used package. The MELF type is cylindrical and leadless and is more commonly used for packaging diodes. The MELF package is discussed in Sec. 2.2.2.

Depending on the manufacturer of the chip resistor, construction and end cap metallization may vary. However, the following description of the internal construction and materials can be considered the most common. Figure 2.3 shows the surface and cross-sectional view of the ceramic chip resistor. In explanation of the surface construction or what is visible to the observer, the largest part of the chip resistor is the ceramic substrate material, which provides attachment and support for the other parts of the chip resistor. On both ends of the chip resistor are metallized caps, which are sometimes called wraparound terminations or end caps. These end caps provide an electrical interconnection between the ends of the resistor element as well as a solderable surface for circuit board attachment. Some end cap designs do not wrap around the part; instead, the metallizations are only on the side where the resistive film is applied. These types are not recommended since the part has to be soldered with the resistive film side down, which leaves an uninspectable solder joint. Making contact with the end caps is the resistive film or element which is overcoated with a fired-on glass passivity or protective film. The protective film eliminates the possibility of solder or conductive adhesive, as the case may be, from leaching onto, or coming in direct contact with, the resistor element, causing resistance change. The protective film also eliminates the possibility of resistance shifts of high values in humid conditions and allows the chip to be safely mounted upside down (resistor film side down), which is sometimes unavoidable when bulk resistors are fed via vibratory feeders.

Reviewing the cross-sectional view of the chip resistor will let us understand the interconstruction and material design of the component (numbers in parentheses refer to Fig. 2.3). As mentioned previously, the chip resistor has a ceramic body (1). The ceramic material is almost always composed of 96 percent grade aluminum oxide (Al_2O_3), which is considered a high-purity material when a maximum of 4 percent of other materials or impurities is present. Higher concentrations of impurities can cause a current flow or change in resistance and poor adhesion of metallization upon reflow soldering.

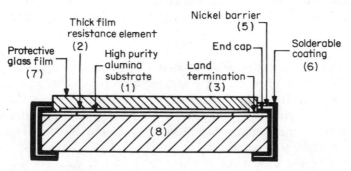

Figure 2.3 Surface and cross-sectional views of the chip resistor.

Laying on top of the body is the resistive film (2) composed of ruthenium dioxide (RuO_2), which extends onto the land terminations at either end of the resistive film. The land terminations (3) act as a connecting surface for both the resistive film and end caps. The end caps (4) are provided as external connections which can be composed of palladium-silver, gold, palladium-silver-platinum, silver, or several other types or combinations of metals.

Although not always available from some manufacturers, a nickel barrier (5) can be applied over the end caps to prevent "leaching" from tin-lead or tin-lead-silver solders. If a nickel barrier is present, a tin or tin-lead protective coating (6) is applied over the nickel to preserve its solderability. This type of end cap is preferred over all others. A glass film (7) composed of lead borosilicate is used to protect the resistive element and is normally laser trimmed. Trimming allows the manufacturer to bring the resistance value to within the required resistance tolerances, which are typically 20, 10, 5, 2, 1, 0.5, 0.25, and 0.1 percent of the normal value.

2.1.3 Resistor networks and variable resistors

For added flexibility in circuit design, resistor networks and variable resistors are available from various sources in surface mount packaging. The *small outline* (SO) package is typically used for resistor networks. The *plastic leaded chip carrier* (PLCC) package is also used but is less common. Both of these packages are described in Sec. 3.1 since they are the dominating package types for integrated circuits.

Surface mount variable resistors or potentiometers are available in numerous types of unstandardized package designs. The most commonly used design is illustrated in Fig. 2.4. For inspectability and reworkability, it is recommended to choose those package designs which have lead or leadless configurations that are easily observed and located so as to make solder rework simple. Also, one must be con-

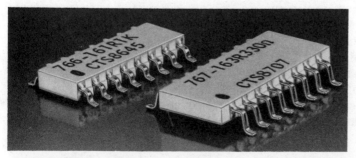

Figure 2.4   The more commonly used surface mount resistor network.

cerned as to whether or not the manufacturer has constructed their part with materials that can withstand the manufacturing assembly processes.

2.1.4 Electrical capacitance

The capacitor is a circuit element that stores electric charges. In its most elementary form, a capacitor consists of two parallel conducting plates separated by an insulator, such as air, glass, barium titanate, or mica. When the plates are connected across the terminals of a battery, electrons start to flow away from the negative terminal, but the charges cannot cross the insulator between the plates. This causes the electrons to "pile up" at the plate connected to the negative terminal. The excess negative charges produce an electric field which causes a repulsion of electrons from the positive plate. If the applied voltage across the capacitor is increased, the stored charge in the capacitor also increases.

A mathematical relationship between the charge and the voltage across the capacitor can be given as

$$Q = CV$$

or

$$C = \frac{Q}{V} = \frac{\text{coulombs}}{\text{volt}} = \text{farads}$$

Here C, the capacitance, is the measure of the charge capacity of the capacitor, Q the electrical charge, and V the applied voltage. The value of capacitance is defined as one farad when the voltage across the capacitor is one volt, and a charging current of one ampere flows for one second. Because the farad is a very large unit of measure and is not encountered in most typical applications, fractions of the farad are commonly used. The smallest fraction of the farad is the picofarad (pF), which equals 1×10^{-12} F.

Although there are many uses for capacitors in electronic systems other than storing an electric charge, understanding the factors which affect capacitance in such a component will allow us to further understand this component. Let us first calculate the capacitance value of the single plate capacitor. For any given voltage, the capacitance value of a single plate component is directly proportional to the geometry and dielectric constant of the component and can be given by the following equation:

$$C = \frac{KA}{ft}$$

where C = capacitance
   K = dielectric constant of dielectric
   A = area of electrode or conductive plate
   t = thickness of dielectric
   f = conversion factor of 4.452 (English unit) or 11.1 (metric unit)

Obviously, greater capacitance value can be achieved by increasing the electrode area, decreasing the dielectric thickness, or increasing the dielectric constants. Because it is physically impractical to increase the area in a single plate component, the concept of stacking capacitors in a parallel array, much like producing a multilayer printed circuit board, was conceived to produce an effective component with more capacitance per unit volume. This type of multilayer capacitor or monolithic capacitor has caused a change in this simple equation for calculating the capacitance of a single plate component, shown as follows:

$$C = \frac{KAN}{ft}$$

where A = total sum of electrode area overlap
   t = reduced dielectric thickness
   N = total number of dielectric layers

2.1.4.1 Monolithic ceramic chip capacitors. The high reliability of monolithic ceramic chip capacitors has a proven history in the military and aerospace industries. They are also extensively used in electronic systems that require a high degree of reliability. Although these are true statements, chip capacitors, which are used in large quantities for commercial applications, have caused a landslide of problems due to the lack of understanding of the materials and processes used for this component.

Like resistors, there are both military and commercial types of capacitors. The EIA RS-198 governs commercial usage and the MIL-C-55681 military and aerospace usage. As specified by EIA RS-198, there are eight standard types of chip capacitors. Figure 2.5 lists their physical dimensions, and Fig. 2.10 lists their capacitance ranges with respect to the type of dielectric.

From a commercial availability standpoint, 15 or more different styles of capacitors can be obtained, including a wide range of capacitor values and body sizes. The most used standard commercial capacitor is the 1206 style. As with resistors, the 1206 chip capacitors are preferred because most problems occur with either the smaller or larger styles.

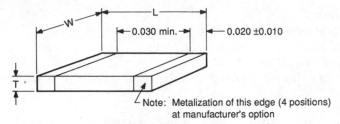

Note: Metalization of this edge (4 positions)
at manufacturer's option

Figure 1

Dimensions* (no solder)**

Style	L		W		T	
	Nom.	Tot.	Nom.	Tot.	Min.	Max.
CC0504	0.050	±0.010	0.040	±0.010	0.020	0.040
CC0805	0.080	±0.010	0.050	±0.010	0.020	0.055
CC1805	0.180	±0.010	0.050	±0.010	0.020	0.055
CC1808	0.180	±0.010	0.080	±0.010	0.020	0.080
CC2225	0.225	±0.015	0.250	±0.020	0.020	0.750
***CC1005	0.100	±0.010 −0.015	0.050	±0.010	0.020	0.050
***CC1210	0.125	±0.015	0.100	±0.010	0.020	0.060
***CC1812	0.180	±0.015	0.125	±0.015	0.020	0.075

*   Dimension in inches.
**   For solder coated terminations add 0.005 to + tolerance on L,
     W maximum T.
*** New

Figure 2.5 EIA's chip capacitor dimensional and style standards
as specified in RS-198C.

2.1.4.2 Chip capacitor construction and materials. Manufacturers of
monolithic chip capacitors produce these multilayer components by
building up successive deposits of dielectric and conductive (electrode)
materials. Figure 2.6 shows the 14 different areas of the chip capacitor. Of these, the terminations and dielectric materials, which are discussed next, are the most critical to the user.

2.1.4.3 Chip capacitor terminations. The chip capacitor terminations,
or end caps, consist of metal-frit compounds which are fused to the capacitor's body to effect an electrical connection between the internal
electrodes and the printed circuit board. The standard end cap metallization is silver-palladium with typical concentrations of 80 percent
silver and 20 percent palladium. Unless electrically conductive adhesives are used as the joining material instead of tin-lead solders, the
silver-palladium end caps are not recommended due to the rapid dissolution of this material with molten solder and the resulting forma-

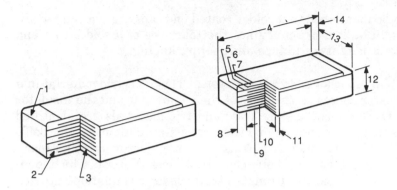

1. Termination or end cap	8. End margin
2. Dielectrode	9. Base layer
3. Electrode	10. Shim (active dielectric layer)
4. Chip length	11. Side margin
5. "A" electrode print	12. Chip thickness
6. "B" electrode print	13. Chip width
7. Cap (topping layer)	14. Termination width

Figure 2.6 The construction of the multilayer chip capacitor.

tion of brittle silver-tin intermetallics. Other end cap materials such as silver and gold have the same dissolution or leaching effects.

The preferred end cap construction and materials, as shown in Fig. 2.7, consist of a silver or silver-palladium inner layer, a nickel or copper barrier layer, and a solder coat or electrolytic tin-lead outer layer. With this end cap construction, the nickel or copper barrier layer, which has a high melting point, prevents the leaching that can occur during the soldering process by not allowing the solder to penetrate to the silver or silver-palladium inner layer. If not protected from oxidation, the barrier metal can cause severe solderability problems and therefore is covered with a solder coat or plated metal such as tin-lead.

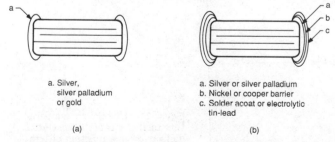

a. Silver,
   silver palladium
   or gold

(a)

a. Silver or silver palladium
b. Nickel or cooper barrier
c. Solder acoat or electrolytic
   tin-lead

(b)

Figure 2.7 End cap metallization constructions for the multilayer chip capacitor. (a) Not preferred; (b) preferred.

It should be mentioned that solder-coated end caps are typically available only in bulk packaging since the tolerance of a solder coat end cap does not lend itself to tape-and-reel packaging.

2.1.4.4   Chip capacitor dielectric materials.   The dielectric material in a chip capacitor serves as the body of the component and the insulator between the electrode layers. The dielectric materials are identified and classified in the industry by their capacitance temperature coefficient (TC). The two basic groups used in manufacturing chip capacitors are identified as class I and class II. A class III exists which identifies the reduced barium titanate barrier layer formulations utilized in manufacturing conventional disk capacitors.

Class I dielectrics.   Class I dielectrics, which are nonferroelectric, display the most stable capacitance characteristics, and are primarily composed of titanium oxide (TiO_2), and have dielectric constants under 150 and a lower capacitance than the class II dielectrics. The most common and stable class I dielectric has been given the designation of COG or NPO, the latter being the military designation. A subgroup of class I dielectrics has a small addition of ferroelectric dielectrics that provide dielectric constants ranging up to 500 or otherwise higher capacitance values than the COG (NPO) dielectric. These subgroup class I dielectrics have negligible or no change in capacitance or dielectric loss with voltage or frequency and predictable linear behavior with temperatures within prescribed tolerances. The subgroup class I dielectrics are designated R2G, S2H, U2H, T3K, and M7G. The linear changes in temperature coefficients of these class I dielectrics are shown in Fig. 2.8.

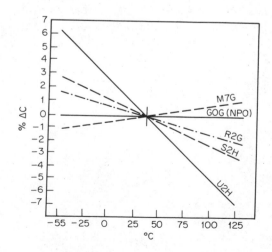

Figure 2.8 Change in temperature coefficients of linear, class I dielectrics as a function of temperature change.

Class II dielectrics. The class II dielectrics, which are typically composed of barium titanate ($BaTiO_3$) and other types of ferroelectric additives, have nonlinear or less stable properties; however, they offer much higher capacitance values at lower costs. The diverse range of properties of class II dielectrics requires organizing them in two categories according to their temperature characteristics. These two categories are designated the stable mid-K class II and the high-K class II dielectrics. The mid-K class II dielectrics display a maximum temperature coefficient of ±15 percent for a 25°C reference over the temperature range of −55 to +125°C. These dielectrics typically have dielectric constants in the 600 to 2500 range and meet EIA RS-198 X7R characteristics. The military specification MIL-C-55681 also defines a mid-K dielectric and has a designation of BX.

The high-K class II dielectrics display dielectric constants from 4000 to 1500 with very steep temperature coefficients, as shown in Fig. 2.9. The most common of these dielectric types have been designated Z5U and Y5V. The Z5U dielectric capacitance ranges are three to five times larger than the X7R dielectric, and the Y5V dielectric offers capacitance ranges greater than the Z5U at a sacrifice of some electrical stability.

2.1.4.5 Capacitance value ranges. Depending on the type of dielectric material used, chip capacitors can be obtained with various capacitance ranges and tolerances. Typical voltage ratings are 100 and 50 V dc. Figure 2.10 shows the capacitance ranges for the standardized styles per RS-198C with the NPO, X7R, and Z5U dielectrics. The smallest standardized value of 1.0 pF incorporates the NPO dielectric,

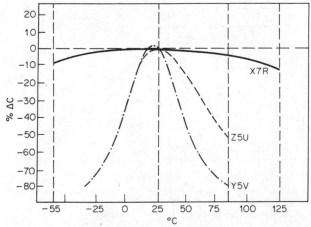

Figure 2.9 Change in temperature coefficients of nonlinear, class II dielectrics as a function of temperature change.

Style	Capacitance range	
	100 Vdc	50 Vdc
CC0504	4.7 to 150 pF	16 to 220 pF
CC0805	4.7 to 220 pF	240 to 680 pF
CC1805	10 to 680 pF	750 to 1500 pF
CC1808	220 to 2200 pF	2400 to 3300 pF
CC2225	1000 pF to 0.01 µF	0.011 to 0.018 µF
CC1005	4.7 to 220 pF	200 to 1000 pF
CC1210	220 tro 820 pF	910 to 2700 pF
CC1812	750 to 1600 pF	1800 to 4700 pF

(a)

Style	Capacitance range	
	100 Vdc	50 Vdc
CC0504	4.7 to 1500 pF	1800 to 4700 pF
CC0805	4.7 to 4700 pF	5600 pF to 0.015 µF
CC1805	220 pF to 0.01 µF	0.012 to 0.033 µF
CC1808	1500 pF to 0.033 µF	0.038 to 0.10 µF
CC2225	0.033 to 0.15 µF	0.18 to 0.47 µF
CC1005	47 to 6800 pF	8200 pF to 0.022 µF
CC1210	1500 to 0.033 µF	0.039 to 0.10 µF
CC1812	0.039 to 0.082 µF	0.10 to 0.18 µF

(b)

Style	Capacitance range	
	100 Vdc	50 Vdc
CC0504	1000 to 2200 pF	2700 pF to 0.01 µF
CC0805	330 pF to 0.01 µF	0.015 to 0.033 µF
CC1805	2200 pF to 0.022 µF	0.033 to 0.10 µF
CC1808	0.022 to 0.068 µF	0.10 to 0.22 µF
CC2225	0.068 to 0.33 µF	0.47 to 1.0 µF
CC1005	4700 pF to 0.01 µF	0.015 to 0.047 µF
CC1210	0.022 to 0.047 µF	0.068 to 0.22 µF
CC1812	0.10 µF	0.15 to 0.47 µF

(c)

Figure 2.10 Capacitance value ranges for the multilayer chip capacitor per EIA RS-198C. (*a*) Class I, NPO dielectric; (*b*) class II, X7R dielectric; (*c*) class II, Z5U dielectric.

whereas the highest standardized value of 1.0 μF incorporates the Z5U dielectric. Higher capacitance values can be obtained with the monolithic construction; however, the costs become noncompetitive with the costs of the tantalum capacitors at the present time.

2.1.5 Tantalum chip capacitors

The tantalum chip capacitor is another type of SMD which can perform the same type of functions as the monolithic chip capacitor. The tantalum chip capacitor, which differs from the monolithic type in both construction and materials, is used where large capacitance and stability of characteristics with temperature and voltage are required. Because tantalum chip capacitors are normally used when there are higher capacitance requirements, these components can be substantially larger in size than monolithic chip capacitors.

The domestic need for tantalum chip capacitors had been for low-volume, specialized applications where product performance, not product standardization, was the main concern. Due to this, these types of capacitors are available in multiple configurations with little agreement between manufacturers as to capacitance value–voltage rating offerings by size or geometry. What standards did exist revolved around MIL-C-55365; however, each manufacturer's configuration was recognized via a unique style sheet.

Now that industrial estimates show an explosive growth in the use of tantalum chip capacitors, the past nonstandardization of this component cannot be tolerated. To accomplish the task of industrial standardization, an EIA work group has established a set of standards in the EIA RC-228B specification. RC-228B specifies two types of tantalum chip capacitors, the unencapsulated and the molded. Both types have five different case sizes with case A the smallest and case E the largest. Figures 2.11 and 2.12 provide the case dimensions for the unencapsulated and molded tantalum chip capacitors.

As specified by RS-228B, the standard voltage ratings and capacitance ranges vary depending on the type of package and case size. Review RS-228B for this information along with specified maximum dc leakage and dissipation factor at specific temperatures for the unencapsulated and molded tantalum chip capacitors.

2.1.5.1 Tantalum chip capacitor construction and materials.
Tantalum chip capacitors are available in a variety of constructions and materials; however, the most commonly used are the unencapsulated, molded, and conformally coated types. The unencapsulated type, as shown in Fig. 2.13, consists of solder-coated nickel terminals which are welded to an external anode post and soldered to the cathode's ca-

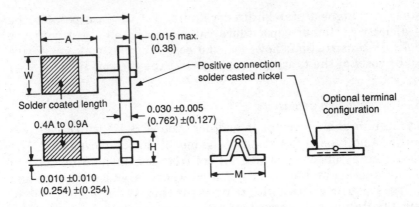

Case size	W		L		H max.		A		M	
	Inches ±0.015	mm ±0.38	Inches	mm	Inches	mm	Inches	mm	Inches ±0.010	mm ±0.254
A	0.085	2.16	0.165 ±0.015	4.19 ±0.38	0.070	1.78	0.065 ±0.015	1.65 ±0.38	0.090	2.29
B	0.105	2.67	0.195 ±0.015	4.95 ±0.38	0.070	1.78	0.105 ±0.015	2.67 ±0.38	0.140	3.56
C	0.145	3.68	0.235 ±0.015	5.97 ±0.38	0.070	1.78	0.145 ±0.015	3.68 ±0.38	0.150	3.81
D	0.150	3.81	0.290 ±0.020	7.37 ±0.51	0.100	2.54	0.180 ±0.020	4.57 ±0.51	0.150	3.81
E	0.155	3.94	0.290 ±0.020	7.37 ±0.51	0.150	3.81	0.190 ±0.020	4.83 ±0.51	0.150	3.81

Figure 2.11 EIA standardized case dimensions for unencapsulated tantalum chip capacitors as specified in RS-228B.

pacitive elements. The capacitive element is based on established porous tantalum anode technology where the anode is sintered and a tantalum pentoxide dielectric layer is formed electrochemically. Using a pyrolytic process, a manganese dioxide coating is then added as the cathode electrode capacitor element. This component is usually the lowest priced tantalum capacitor since it is not encapsulated or sealed from the environment. Also, due to this component's irregular geometry, automatic placement of this type may incur some difficulties.

The molded tantalum chip capacitor, as shown in Fig. 2.14, is simply an unencapsulated type that is potted with a moldable resin such as epoxy. The leads, which are normally solder-coated with a nickel or a copper-nickel-tin metal sheet, are welded to the anode post and sol-

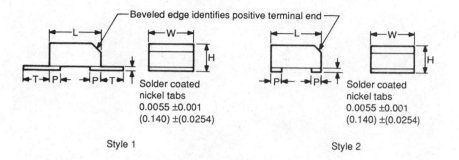

Case size	Style no.	L		W		H		P		T	
		Inches ±0.005	mm ±0.127	Inches ±0.010	mm ±0.254	Inches ±0.005	mm ±0.127	Inches ±0.010	mm ±0.254	Inches ±0.010	mm ±0.254
A	1	0.180	4.57	0.100	2.54	0.070	1.78	0.030	0.762	0.150	3.81
A	2	0.180	4.57	0.100	2.54	0.070	1.78	0.030	0.762	No tabs	
B	1	0.180	4.57	0.100	2.54	0.100	2.54	0.030	0.762	0.150	3.81
B	2	0.180	4.57	0.100	2.54	0.100	2.54	0.030	0.762	No tabs	
C	1	0.320	8.13	0.180	4.57	0.070	1.78	0.050	1.27	0.110	2.80
C	2	0.320	8.13	0.180	4.57	0.070	1.78	0.050	1.27	No tabs	
D	1	0.320	8.13	0.180	4.57	0.100	2.54	0.050	1.27	0.110	2.80
D	2	0.320	8.13	0.180	4.57	0.100	2.54	0.050	1.27	No tabs	
E	1	0.320	8.13	0.180	4.57	0.195	4.95	0.050	1.27	0.110	2.80
E	2	0.320	8.13	0.180	4.57	0.195	4.95	0.050	1.27	No tabs	

Figure 2.12 EIA standardized case dimensions for molded tantalum chip capacitors as specified in RS-228B.

dered to the capacitor element. The preferred lead configuration wraps around the component so that a solder fillet can be easily observed.

Although there are some construction differences, the conformally coated tantalum chip capacitors are made of materials similar to those found in the other two constructions. As shown in Fig. 2.15a, there are military and industrial conformally coated designs. The only basic difference between these two types is the construction of the end caps. The military type has end caps that wrap completely around the component with one side partially conformally coated. For soldering concerns, this makes this side the top of the component. The end cap de-

Figure 2.13 Five unencapsulated tantalum chip capacitors among various other types of SMDs.

Figure 2.14 The molded tantalum chip capacitor. (*Courtesy Hilton Industries, Sarasota, Florida.*)

sign on the industrial conformally coated tantalum chip capacitor does not completely wrap around the component, as shown in Fig. 2.15*b*. In both cases, the preferred mounting position is such that the exposed metallized sides make contact with the printed circuit board lands or pads.

A closer look at the material and construction of the conformally coated chip capacitor can be viewed in Fig. 2.16. As in the molded type, the center consists of an anode composed of porous tantalum. The anode is covered with electrolytically formed tantalum pentoxide which is the dielectric layer. Making contact with the dielectric layer is the electrolyte, manganese dioxide. The electrolyte is covered with

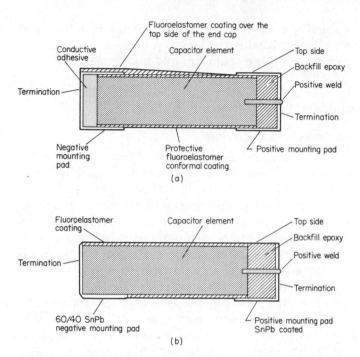

Figure 2.15 Internal construction of the encapsulated tantalum chip capacitor. (a) Military construction; (b) industrial construction. (*Courtesy Mepco/Electra.*)

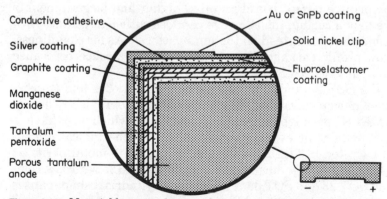

Figure 2.16 Material layers and compositions composed in the conformally coated tantalum chip capacitor. (*Courtesy Mepco/Electra.*)

three layers of conductive materials consisting of graphite, silver, and conductive adhesive. These conductive layers provide a low resistive path and connection between the tantalum and manganese dioxide.

2.2 Discrete Components

2.2.1 Small outline diodes and transistors

The small outline diodes and transistors designed for surface mounting, commonly called SOT 23's and SOT 89's, were developed by North American Philips Corporation in the early 1970s. Compared to conventional packaging types, packaging semiconductor diodes and transistors in surface mount constructions has drastically reduced the space these components occupy on the printed circuit board. Furthermore, these components can be pretested, are easy to position for automatic placement, have good thermal characteristics, and are reliable enough to withstand the various reflow soldering techniques.

The SOT 23, which is the most commonly used package, and the SOT 89 have typically standard dimensions, as illustrated in Figs. 2.17 and 2.18. The standoff height clearances, as shown in Fig. 2.19, for the SOT 23 are defined as low, medium, and high. When choosing the standoff height clearance, one must first know the type of reflow process to be used. For vapor condensation or infrared reflow soldering, any one of the three standoff height clearances is satisfactory when bonding adhesives are not used; whereas in dual wave soldering, the low-profile design will provide the best conditions when bonding this component to the board. Bonding medium- and high-profile packages to the printed circuit board requires higher and larger amounts of adhesive. Upon curing, the adhesive can lose sufficient contact with the component body or spread on the land patterns. As for cleanliness, with the low-profile package, this should never be a concern because of the small surface area involved.

Besides the SOT 23 and SOT 89 packaging for diodes and transistors, another newer packaging exists, called the metal electrode face bonding (MELF) package. The MELF package, which can have a semiconductor diode or even a resistor film, has the typical dimensions as illustrated in Fig. 2.20. Providing a glass-to-metal hermetic seal, this type of package has higher reliability and lower costs compared to the SOT 23 and SOT 89. The MELF's cylindrical shape causes possible automatic placement difficulties, so it is not the most commonly used type of component. However, when used in significant quantities and assemblies, the lower cost of the MELF package will compel placement machine manufacturers to accommodate to the ease of automatic placement of this package.

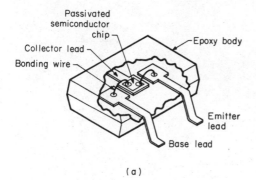

(a)

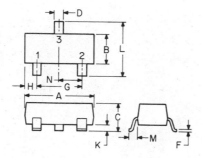

Dim.	Millimeters		Inches	
	Min.	Max.	Min.	Max.
A	2.80	3.05	0.110	0.120
B	1.20	1.40	0.047	0.055
C	0.85	1.20	0.033	0.047
D	0.37	0.45	0.015	0.0179
F	0.085	0.132	0.003	0.0052
G	1.78	2.04	0.070	0.080
H	0.51	0.60	0.020	0.024
K	0.10	0.25	0.004	0.010
L	2.10	2.50	0.083	0.098
M	0.45	0.60	0.018	0.024
N	0.89	1.02	0.035	0.040

Figure 2.17 The SOT 23 surface mount package. (*a*) Internal construction; (*b*) typical case dimensions. Maximum die size 30 mil × 30 mil.

(b)

2.2.2 Small outline diodes and transistor construction and materials

The constructions for the SOT 23 and SOT 89 packages are illustrated in Figs. 2.17 and 2.18, respectively. Both packages typically have either silicon or epoxy bodies with lead frame materials composed of one

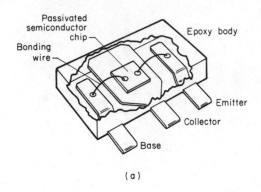

(a)

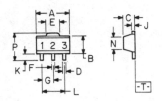

Dim.	Millimeters		Inches	
	Min.	Max.	Min.	Max.
A	4.40	4.60	0.174	0.181
B	2.29	2.60	0.091	0.102
C	1.40	1.60	0.056	0.062
D	0.36	0.48	0.015	0.018
E	1.62	1.80	0. 064	0.070
F	0.44	0.53	0.018	0.020
G	1.50 BSC		0.059 BSC	
J	0.35	0.44	0.014	0.017
K	0.80	1.04	0.032	0.040
L	· 3.00 BSC		0.118 BSC	
N	2.04	2.28	0.081	0.089
P	3.94	4.25	0.156	0.167

Figure 2.18 The SOT 89 surface mount package. (*a*) Internal construction; (*b*) typical case dimensions. Maximum die size 60 mil × 60 mil.

(b)

of the many types of copper alloys which are tin-lead-plated. The standard die metallization is aluminum. Gold wire is used when the highest degree of reliability is required, and silver wire where a lesser degree of reliability and cost is needed. The SOT 89 package is capable of handling higher wattages than the SOT 23 package since the extra large tab area for chip attachment acts as an internal head sink to transfer the heat out of the junction area.

The internal construction of the MELF package is illustrated in Fig. 2.20*a*. With only two outside contact areas, this package's simplistic

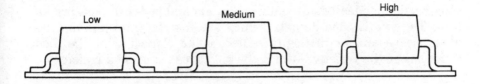

Stand off height clearance	Millimeters		Inches	
	Min.	Max.	Min.	Max.
High	0.100	0.250	0.0040	0.0100
Medium	0.076	0.130	0.0030	0.0050
Low	0.010	0.100	0.0005	0.0040

Figure 2.19 SOT 23's various standoff height clearances.

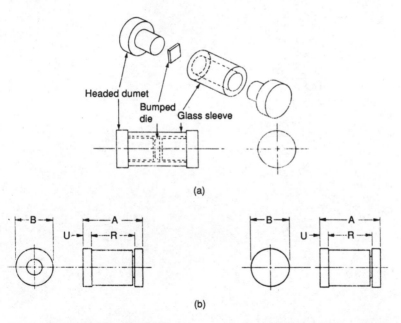

(a)

(b)

Dim	Millimeters		Inches	
	Min.	Max.	Min.	Max.
A	4.80	5.20	0.189	0.205
B	2.44	2.54	0.096	0.100
R	3.71	4.59	0.146	0.181
U	0.36	0.50	0.014	0.020

Dim	Millimeters		Inches	
	Min.	Max.	Min.	Max.
A	3.30	3.70	0.130	0.145
B	1.60	1.70	0.063	0.067
R	2.49	2.59	0.098	0.102
U	0.41	0.55	0.016	0.022

Figure 2.20 The MELF surface mount package. (*a*) Internal construction; (*b*) typical case dimensions.

construction is ideally suited for high-yield and low-cost mass production. The two headed dumets, which are inserted into a cylindrical glass sleeve and hermetically sealed, make contact with the diode functions or the resistive film if the package is used as a resistor.

Bibliography

Arcidy, Michael, "Maximizing Chip Resistor Performance and Reliability for Surface Mount Applications," *Evaluation Engineering,* September 1986.

Galliath, Andre P., *Ceramic Chip Capacitors (Technical Brochure),* Novacap, Inc., Burbank, California.

Meeldijk, Victor, "Latent Defects Result in Multilayer Ceramic Capacitor Failures," *Electronic Packaging and Production,* January 1984.

Winard, Harold, "Focus on Surface Mounted Electrolytic Capacitors," *Electronic Design,* August 7, 1986.

Surface Mount Directory, 1987–88, Information, Inc., New York, New York.

Active Surface Mount Components

3.1 Surface Mount Integrated Circuit Components

Surface mount integrated circuit (IC) components come in a variety of package designs, materials, leaded or leadless configurations, and with from 8 to higher than 340 input/outputs (I/Os) per package. The choice of a particular package design and environment can drastically affect the manufacturability, reliability, and cost of the assembly. This is why it is of the utmost importance that not only the component engineer, but also the component purchaser and design engineer understand the relationship of manufacturing, reliability, and cost when choosing a particular surface mount package for their application. The obvious reason behind this is that, unlike in the conventional through-hole technology, the purchasing and design staff can totally affect a company's success or failure in using and manufacturing surface mount assemblies.

It is the intent of this section to provide basic information on the most common surface mount IC packages and to address in a non-self-serving way the latent or potential problems, which are not usually pointed out by the component manufacturer, associated with the particular package design.

3.1.1 Small outline integrated circuit

The small outline integrated circuit package, or SOIC, which was introduced in Europe by N. V. Phillips in the early 1970s, has promised considerable advantages in packaging density, reliability, and manufacturing assembly. Although dimensional standardization of these components between component manufacturers has progressed in very controversial stages, today three separate standards exist: (1) The Joint Electronic Device Engineering Council (JEDEC), which most North American component manufacturers conform to, (2) the International Electrotechnical Commission (IEC), and (3) the Electronic Industries Association (EIA). These standards include both gull-wing (SOIC) and the J-leaded (SOJ) configurations with leads on only two opposite sides of the package.

Prior to discussing the variations among these three standards, let us first consider the two major conditions that will affect the assembly's reliability, manufacturability, and inspectability: the heel-to-opposite-heel distance and the lead type. In order for the designer to provide the proper amount of land extension behind the heel, the heel-to-opposite-heel distance must be known. Unfortunately, the current standards do not provide this information directly, nor do most component manufacturers in their own specifications. It is therefore necessary for the designer to calculate this dimension and specify the minimum and maximum dimensions on the purchase order or procurement specification.

The lead type, whether gull-wing or J configuration, should be an important consideration. First of all, let's clarify the real reason between the existence of these two types of lead configurations. At the component manufacturing level, the J-lead configuration is preferred due to the fewer requirements imposed on the manufacturer to provide adequate controls to prevent lead damage.

On the other hand, some component manufacturers have good handling techniques and can produce the gull-wing configurations along with providing to their customers the added values associated with this lead. These added values of the gull-wing over the J-lead configuration may still be controversial; however, the author's opinions are based on the user's needs, not the component manufacturer's. Simply stated, the gull wing's advantages over the J lead are numerous. These include the added stress relief or compliancy due to the lead extension design and bend radius from the component body which adds reliability, the ease of reworkability, the ease of inspectability, and as a result of these advantages, higher manufacturing yields. Furthermore, one might say that the J-lead configuration or SOJ package takes up less space on the printed circuit board than the gull-wing package and, therefore, is the preferred type. The author would concur with this statement, provided that the end user understands that when high density is mandatory, one should only expect a gain in density and a possible loss or total loss in visual inspection, test, and rework capabilities.

3.1.2 SOIC standardization

As mentioned, three separate standardization efforts have occurred for the SOIC package under the organizations of JEDEC, IEC, and EIA. Basically there are 0.300- and 0.150-in. wide-body gull-wing SOIC-leaded packages and 0.300- and 0.400-in. wide-body SOJ-leaded packages. Although standardization should imply uniformity, this is clearly not the case when comparing the individual standardization efforts of JEDEC, IEC, and EIA. As reported by E. S. Codon of ITT, JEDEC has three proposed SOIC standards and three for SOJ. The IEC has four different prepared standards, and the EIA J has eight proposed standards which have been consolidated into four standard sizes for ease of comparison.

To laboriously detail the differences between these standards is beyond the scope of this chapter; however, since most U.S. component manufacturers have standardized their SOICs to JEDEC, let us take a brief look at what this offers us as users. First of all, we can expect from various U.S. component manufacturers basically the same case outline dimensions and tolerances, which will not adversely affect the user's needs. Contrary to standardization are the wide tolerances of the gull-wing lead profiles for the SOIC, which are in conformance to JEDEC standards. As a result, and as shown in Fig. 3.1, the user can obtain lead profiles in foot lengths from 0.010 to 0.050 in. and foot-length tolerances of as high as 0.034 in. for the same part number. Based on this, one might ask, When will I receive a 0.050-in. foot length for the same part number? Obviously, these tolerances are not acceptable if one is to standardize both the company's printed circuit board designs and manufacturing techniques. Since the standardization organizations have more work to do in order to provide adequate standards, it is recommended that the user measure the actual dimensions and tolerances of the received component and then specify these conditions in the procurement document. In all cases, it will be found that the tolerances measured for case outlines and lead profiles will be consistent over time from the respective manufacturer.

3.1.3 SOIC construction and materials

The SOIC package is constructed much like the plastic dual-in-line (DIP) package with its stamped or etched lead frame, standard IC die, and gold wire bonded to the die pads and lead frame pads. Whether formed according to gull-wing or J-lead configurations, the lead frames are preferably perforated between the lead frame bonding pads and just inside the case edge, as shown in Fig. 3.2. These perforations

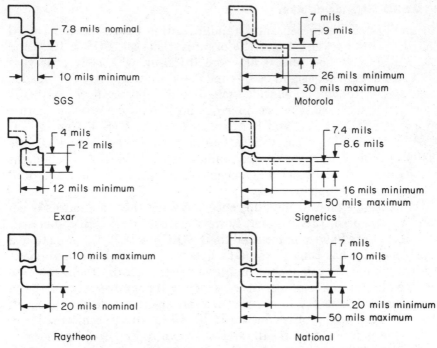

Figure 3.1 Various component manufacturers' lead profile dimensions for SOICs that are all in conformance to JEDEC standards. (*Courtesy D. Brown and Associates.*)

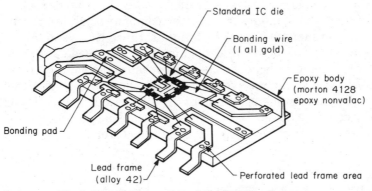

Figure 3.2 The internal construction of the SOIC package.

provide mechanical strength to aid in the adhesion of the resin to the lead frame. The initial designs for these packages normally did not provide these perforations, which caused a loss in resin adhesion to the lead frame especially during the reflow process of this component. This in turn produced openings to the environment from which contaminants such as flux residues, solvents, moisture, etc. could enter and cause component failure.

Not only does a design consideration such as the perforations on the lead frames of the SOIC package affect reliability, but it has also been found that the types of metal alloys the lead frames are composed of can have significant effects on component performance as it relates to long-term reliability when the component is assembled on the printed circuit board. Since the materials for the SOIC packages are normally the same as in other surface mount plastic packages, the concerns relating to these package and lead materials will be discussed in the following sections.

3.1.4   Plastic leaded chip carriers

The plastic leaded chip carrier (PLCC), sometimes called a quad package, is available in both the premolded type and the more common postmolded type. Both types have J leads on all four sides of the package. The J leads may be the free-standing type or the less compliant, and thus less popular, tight-wrapped type, each of which is shown in Fig. 3.3. The PLCC has certainly been a popular choice among users due to the belief that sufficient compliancy, high density, lead coplanarity, and less lead damage are gained by using this component over the other plastic surface mount packages. Although this is not intended to be a court trial between the PLCC and the other plastic gull-wing packages including the SOIC and flat pack (to be discussed later), the user should be concerned that due to board designs, excessive amounts of board warp and twist, reduced visual inspection, and high leaded counts, some applications with the PLCC may produce lower levels of reliability. This, of course, has to do with the individual

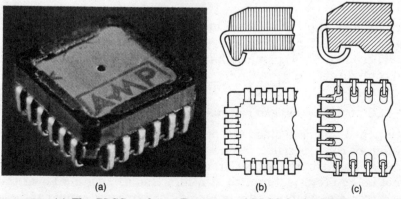

(a)   (b)   (c)

Figure 3.3 (a) The PLCC package. Two types of PLCC lead configurations: (b) The tightly wrapped lead; (c) the free-standing lead.

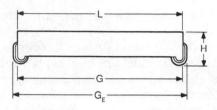

Lead count	L	H	G	G$_E$
20	0.350 ±0.005	0.165 – 0.180	0.320 ±0.010	0.390 ±0.005
28	0.450 ±0.005	0.165 – 0.180	0.420 ±0.010	0.490 ±0.005
44	0.650 ±0.005	0.165 – 0.180	0.620 ±0.010	0.690 ±0.005
52	0.750 ±0.005	0.165 – 0.180	0.720 ±0.010	0.790 ±0.005
68	0.950 ±0.005	0.165 – 0.180	0.920 ±0.010	0.990 ±0.005
84	1.154 ±0.004	0.165 – 0.180	1.120 ±0.010	1.190 ±0.005
124	1.654 ±0.004	0.190 ±0.010	1.620 ±0.010	1.690 ±0.005

Figure 3.4 The JEDEC standardized package dimensions for the PLCC.

solder joints connecting the PLCC to the printed circuit board, not with the actual functional reliability of the component.

The PLCC has been standardized by JEDEC into eight different lead counts, all on 0.050-centerline spacings and square bodies. The dimensions of these packages, to which most manufacturers conform, are illustrated in Fig. 3.4. The lead dimensions from different manufacturers can vary; however, the dimensions of the leads themselves usually fall within the those shown in Fig. 3.5. The thicknesses of the leads are normally 0.006, 0.008, or 0.010 in. and composed of various metal alloys.

Furthermore, the PLCC is not only available in a square package but also in a rectangular package typically intended for memory chips. These packages currently have lead counts from 18 to 32 with J-lead configurations.

3.1.5 PLCC construction and materials

As mentioned, the premolded and postmolded PLCC types exist as the two different constructions. The premolded PLCC consists of a plastic housing or base where the lead frame and die are assembled and capable of being pretested prior to attaching and sealing a lid over the die-lead frame assembly. Although this type enjoyed limited success and promised lower manufacturing and tooling costs, it did not find its

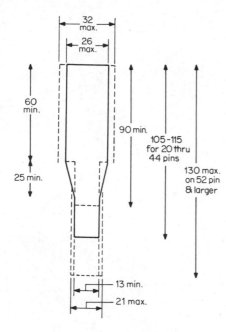

Figure 3.5 Component manufacturers' various designs and dimensions for the J-lead configuration. Note: All dimensions in mils.

way in any significant amount to the user and thus does not warrant any further discussion.

The postmolded PLCC is manufactured in much the same way as the DIP package in that the plastic is molded over the lead frame after the die has been attached via wire bonding or tape automated bonding (TAB). The internal construction of the postmolded PLCC is illustrated in Fig. 3.6. As shown on the J-lead configuration, the identified areas include the metal lead frame, the encapsulant, the bonding wire or tab beam, the die, and the die attach conductive adhesive. Since the die, die attach conductive adhesive, and wire or tab beam are typically the same materials as in the DIP package, only the metal lead frame, lead coatings, and encapsulant will be discussed further; all of these play a vital role in making the PLCC a functionally reliable component.

3.1.5.1 Component encapsulating materials.

The material used to enclose or encapsulate the lead frame and die assembly is usually ceramic, polyetherimide, fused silica, or epoxy resin. The ceramic material, normally beryllium oxide (BeO), is used for high-reliability electronic assemblies such as found in military and aerospace systems. This is the only material which provides hermeticity at levels which pass the fine leak testing. This, of course, is an advantage for this package when hermeticity is a requirement; however, due to the

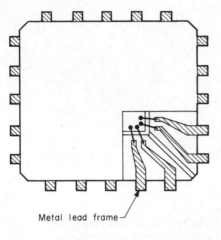

Metal lead frame

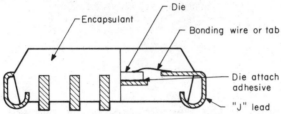

Figure 3.6   The internal construction of the PLCC.

lead configuration, it also provides much more compliancy or stress relief than does the ceramic leadless chip carrier (LCC), which is one of
the common packages for high-density military and aerospace applications. Due to the additional compliancy the J-lead PLCC offers over
the LCC, expensive coefficient of thermal expansion (CTE) substrates,
which are mandatory when using LCCs, are usually unnecessary with
the ceramic J-lead carriers.

The polyetherimide, fused silica, and epoxy resins are used as
encapsulants for the PLCC. The polyetherimide material, which is
only used for the less popular premolded PLCC, is a glass-filled, high-
temperature thermoplastic with excellent mechanical and electrical
properties and low ionic contaminants (below 10 ppm). Unlike the
other plastic encapsulants, this material is sensitive to infrared radiation (IR) between various tested wavelengths. When exposed to IR,
the polyetherimide decomposes into a "bubbling" mass of material.
Simply stated, when soldering PLCCs composed of polyetherimide, IR
reflow does irreversible damage to the component. The fused silica
and epoxy resins, which are thermoset materials, are the most common encapsulants used for plastic packages. Unlike those used in the

earlier stages of development, these materials are generally of low expansion, high purity, and high compression for improved density and resistance to moisture penetration.

3.1.5.2 Lead frame materials . In the early stages of developing plastic surface mount packages, it was thought by most component manufacturers that the same lead frame materials or lead material used in plastic DIP packages would produce satisfactory results with surface mount packages. These materials or alloys, normally Kovar or alloy 42 (iron-nickel), are fairly stiff or hard and therefore resist damage from handling. However, when used for the J configuration, the lead thickness coupled with the wide shoulder design and very short, narrow legs produces a lead with very little compliance.

As a result of these initial problems and concerns, especially with the J-lead configuration, the IEEE Computer Packaging Committee was organized by J. Blade of Interconnection Decision Consulting (Flemington, N.J.). Since 1984 the committee has undergone comparative evaluations of lead materials from various PLCC manufacturers in order to determine each lead material's composition, thickness, hardness, grain orientation, and finish. Their initial findings are illustrated in Fig. 3.7. Review of these data indicates that among these suppliers, three common lead alloy materials are used: the copper alloy C151, the copper-iron alloy C194, and the iron-nickel alloy 42 with compositions as shown in Fig. 3.8.

Also of major concern with some applications are the thermal and electrical conductivities of these metal alloys. Alloy C151 has superior

Supplier	Lead alloy	Hardness	Thickness	Type	Coatings
AMP	C194	115	0.006	Etch	90Sn-10Pb/plate
AT & T	C194	124	0.008	Etch	Au
Hitachi	Alloy 42	166	0.006	Stamp	75Sn-25 Pb/plate
JADE	C151	59	0.010		75Sn-25 Pb/plate
Kijocera	Alloy 42	170	0.006	Etch	Au
LSI	C151	97	0.008	Etch	Sn
National	C194	111	0.010	Stamp	85Sn-15Pb/plate
NCR	C194	85	0.008	Etch	Ag
Signetics	C194	113	0.008	Etch	70Sn-30Pb/solder coat
Texas Instruments	C151	95	0.008	Etch	Sn

Figure 3.7 Lead characteristics from various component manufacturers as analyzed by the IEEE Computer Packaging Committee.

Nominal composition - weight percent					
Alloy	Min. Cu (including Ag)	Fe	Sn	P	Other
151	99.9				0.10 Zirconium
155	99.75			0.06	0.027-0.10 (8-30 oz/ton) Ag 0.08-0.13 Mg
194	97.0	2.35		0.03	0.12 Zn 0.03 Pb max.
195	96.0	1.50	0.6	0.03	0.8 Co
42	←——— TBD ———→			←——— TBD ———→	

Figure 3.8 Compositions of the most common lead frame metal alloys used in surface mount leaded packages.

thermal and electrical conductivities over the other alloys. This alloy, which is mostly copper, has small amounts of zirconium, which gives increased strength and softening resistance to the copper but at the same time shows no significant loss in electrical and thermal conductivity. Certainly the C151 alloy provides most of the necessary requirements for a reliable J-lead configuration. On the other hand, when leads are gull-wing shape, as in the SOIC, adding hardness or stiffness to the lead is necessary to minimize lead damage caused by handling. In this case, metal alloys such as alloy 42 may be preferred without a significant loss in compliancy.

3.1.5.3 Lead coatings and finishes. Lead coatings and finishes, which can also vary between component manufacturers, consist of several tin (Sn)-lead (Pb) plating compositions, tin-lead solder coat, tin, silver (Ag), and gold (Au). With lead coatings, the most important factor is the solderability of the lead. Without a doubt, solderability plays a vital role in reliability and manufacturability and is currently the single major problem with surface mount components. Although discussed and written about frequently, lead coplanarity and damage during handling have not been observed at any significant level even under high-volume usage. Solderability, therefore, should be our major focus and is discussed in detail in Chap. 7. But let us answer the question, Which lead coating provides the best solderability over time, temperature, and aging? During the past 20 years the answer has been the same—*solder coat,* which is sometimes called *hot solder dipped.* Since solder coat is a fused alloy of 50 to 70 percent tin and the remainder lead, this coating forms weak metal oxides over a period of time, unlike the unfused and plated tin and tin-lead finishes. The plated tin and tin-lead coatings can form heavy metal oxides over a relatively short period of time, and these are more difficult to remove during soldering, especially with the common mildly activated fluxes (RMA) in solder pastes. Typically, the gold and silver lead coatings have to be removed prior to soldering due to the formation of brittle intermetallics consisting of gold-tin and silver-tin, respectively.

What has been discussed about lead coatings is a controversial subject among component manufacturers since most of them are currently only willing to provide the user with what is easiest for the manufacturer—the plated lead finish. Due to the seriousness of solderability and the large effect it can have on reliability and yields, some responsible component manufacturers now provide leaded packages with solder-coated leads. If zero defect manufacturing is the goal, solder-coated leads will have to be a mandatory requirement.

3.1.6 Flat packs

Since the 1960s, flat packs, or flat packages, have been the most commonly used surface mount package for military and aerospace applications. This package consists of a ceramic body with gull-wing-type lead configurations which exit the body in either the straight-leaded or spider-leaded version as illustrated in Fig. 3.9. Although this package is considered the "father" of surface mount ICs, it is losing popularity due to handling difficulties and lower packaging density as compared to the LCC and the ceramic PLCC.

Another flat pack has the same basic construction and materials as the PLCC. The leads, which are in gull-wing configuration, exit on all four sides of the package, as shown in Fig. 3.10. The lead centerline spacings normally range between 0.050 and 0.020 in. with I/Os ranging from 16 to 340. Although this plastic flat pack has more reliable lead compliancies when compared with the PLCC, it has not been a favorite package design primarily because most manufacturers are unable to assemble this component with high yields. Certainly the

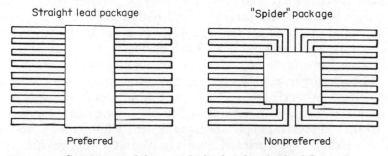

Figure 3.9 Comparison of the straight lead and spider lead flat packs.

Figure 3.10   The gull-winged plastic flat pack.

lower profile of the flat pack, which is approximately half that of the PLCC, is a significant advantage in some applications.

3.1.7 Chip carriers

The variety of chip carriers that exist today started as a result of the hermetic leadless chip carrier development and introduction by 3M in the early 1970s for military and aerospace applications. When compared to the same I/O counts, this leadless version, as shown in Fig. 3.11, offers higher packaging density than any other surface mount IC package. Area reductions to 5:1 or more are possible throughout the complete range of package sizes, which also provide minimum lead frame lengths offering increased high-frequency performances.

The hermetic leadless chip carrier (LCC) is available in both the single-layer, aluminum-metallized (SLAM) type and the three-ceramic-layer type known as the LCC. Each has been standardized by JEDEC with the three-layer LCC the most commonly used package.

The SLAM chip carrier consists of a single-layer-type ceramic substrate made of either beryllium oxide (BeO) or aluminum oxide and a pressed ceramic, cup-shaped lid. Another SLAM chip carrier intended for high-density commercial applications is typically fabricated much like a two-sided, copper-clad printed circuit board but normally with a polyimide epoxy glass substrate. The commercial SLAM package,

Figure 3.11 The hermetic leadless chip carrier (LCC), type C.

sometimes called a plastic leadless chip carrier, uses either a plastic cap or epoxy glob for environmental protection of the die.

The LCC is available in type A, B, C, D, E, and F configurations, as shown in Fig. 3.12. The type C LCC is actually the original version of the chip carrier and the type most commonly used. The internal construction of type C, as illustrated in Fig. 3.13, consists of three separate layers of BeO ceramic used for the base external contact, internal bond pads, and conductors and the seal ring layer made to bond to a ceramic or metal lid. The metallized pads or castellations only extend up to the second layer and the bottom edge of the base–external contact layer. Normally, the base metal throughout the package is tungsten with a nickel barrier and gold-plated coating.

The current JEDEC standards and proposed standards for the LCC, as illustrated in Fig. 3.14, provide 16 to 308 I/O packages on 0.050-, 0.040-, 0.025-, and 0.020-in. centerline spacings. As discussed in Chap. 6, because of the mismatch between the coefficient of thermal expansion (CTE) of leadless type C packages and that of typical substrate materials such as epoxy FR-4 and polyimide, the leadless type C packages are mounted on substrates designed to control CTE. Furthermore, LCCs with I/Os greater than 68 and those with 0.025- and

Figure 3.12 The five types of LCCs: A, B, C, D, E, and F.

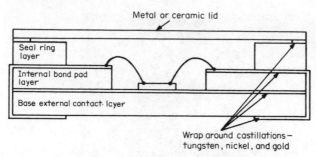

Figure 3.13 The internal construction of the type C LCC.

Overall dimension, inches	Centerline spacing			
	0.050 in	0.040 in	0.025 in	0.020 in
0.235	—	16*	—	—
0.300	16	20	28	36
0.330	—	20*	36	44
0.350	20*	24*	36	48
0.400	24	28	44	60
0.420	—	32*	44	60
0.450	28*	32	52**	68**
0.480	—	40*	60	76
0.560	—	48*	68**	92**
0.650	44*	52	84**	108**
0.720	—	64*	92	124**
0.750	52*	64	100**	124
0.920	—	84*	124	164
0.950	68*	84	132**	164**
1.040	—	96*	148	188
1.150	84*	104	164**	204**
1.350	100*	124	196**	244**
1.650	124*	152	244**	308**
2.050	156*	192	—	—

** Proposed JEDEC standard.
* JEDEC standard.

Figure 3.14 The proposed and standardized JEDEC LCC package dimensions.

0.020-in. centerline spaces require even more exotic substrate material designs and assembly manufacturing techniques.

3.1.8 Leaded chip carriers

The type A and type B LCCs also exist in leaded configurations for direct mounting to the substrate. Type A can also be assembled in a socket but type B cannot. The leaded version of the LCC is normally used to replace or avoid the use of the expensive tailoring expansion substrates by providing adequate stress relief or compliancy between the component's body and that of the common substrate materials.

Between the two leaded types, type A is considered the most reliable and standardized configuration. The type A lead design is of the J configuration and is normally brazed to the top surface of the LCC. The lead base materials for both A and B types are normally the same as found in SOIC packages with either tin, tin-lead, gold-lead platings, or a solder coat.

3.1.8.1 Clip leads and chip carrier mounting components. The type B leaded LCC, which is not socketable, has edge clip leads of various designs soldered to one or both sides of the package. The clip lead assembly to the type B LCC is illustrated in Fig. 3.15 with typical dimensions provided for the clip lead. This particular soldered clip lead design is also popular for commercial hybrid packages; however, under certain conditions this type of clip lead can have low levels of reliability.

Another lead construction for the LCC, again used for compliancy purposes, is called the chip carrier mounting device (CCMD), developed by Raychem Corporation. This lead or component consists of a cylindrical column of solder with a flat copper wire wrapped around it in a helical shape. Each column is soldered to the bottom side of a type C LCC so that it is directly under the package and perpendicular to it when assembled on the substrate material.

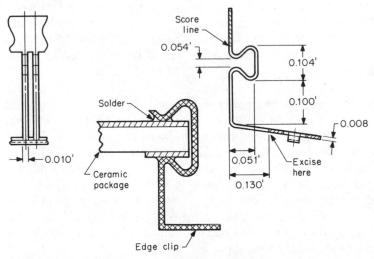

Figure 3.15 The common edge clip lead used for direct mounting type B LCCs.

3.1.9 High I/O count leaded components

High I/O count leaded surface mount components in excess of 100 pins have been incorporated in military and high-reliability electronics since the early 1970s. They are normally used when high packaging density is required, which is a typical condition for very large scale integrated circuits (VLSI) and very high scale integrated circuits (VHSIC). Ranging in a variety of pin counts, Fig. 3.16 illustrates three different high I/O count packages with 132, 256, and 340 pins per package with lead centerline spacings on 0.020 in. Of course, there are other common packages with a variety of lead counts with 0.020- and 0.025-in. lead centerline spacings.

3.1.9.1 Standardization for high I/O count leaded components.   Although the EIA, JEDEC, and IEC have not yet approved standards for high I/O count and fine centerline space SMDs, it is a major concern from the user's standpoint that those involved in developing the standards actually understand the effects the component design and materials have on the ability to assemble such high I/O count components. In some cases this has not been true for even the simplest SMDs.

First of all, leadless high I/O count components with centerline spacings less than 0.040 in. should not be developed into standard components. The main reason for this is that with the LCC, as the I/O count increases, the reliability decreases. When the I/O count for a leadless component exceeds approximately 64, it is likely that such a component and certainly larger ones will impose extraordinary problems such as cleanliness, lack of compliancy or stress relief to overcome printed circuit board warp and twist, low soldering yields, and very low tolerance to vibration fatigue. The only exception in favor of a leadless high I/O count package would be if it were intended for a leaded socket.

3.1.9.2 High I/O count leaded package design.   Although a variety of package designs for high I/O count components exist, this has not been a major concern during high-volume manufacturing assembly of these components. However, a package design utilizing the so-called lead-protecting bumpers developed by AT&T would offer several advantages over the typical square-body packages. The bumpers, which protrude from each corner of the package, go beyond the ends of the leads and can protect the leads from handling and packaging damage. Although the bumpers will provide more lead protection, they alone are not adequate to minimize lead damage.

The major concern in high I/O count component design is the construction of the lead. As the package size or lead count increases, the

(a)

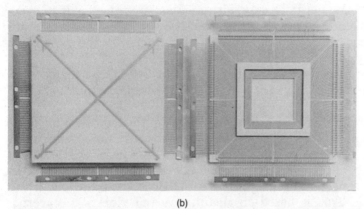

(b)

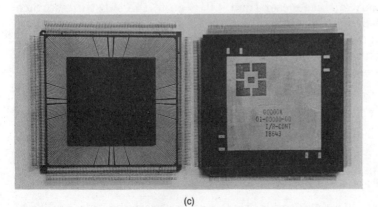

(c)

Figure 3.16 Three types of high I/O count SMDs. (*a*) 132 pins; (*b*) 256 pins; (*c*) 340 pins.

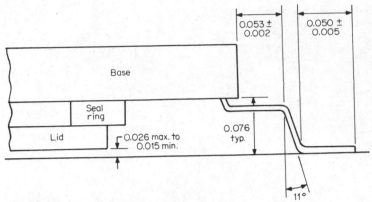

Figure 3.17 Manufacturable package and lead design for high I/O count SMDs.

compliancy of the package must also increase in order for it to overcome CTE mismatch, warp and twist of the substrate, and vibration stresses. A lead design which has worked well for both the component and assembly manufacturer is illustrated in Fig. 3.17. This lead design has been incorporated in the 256- and 340-I/O packages with 0.020-in. centerlines and is utilized in high-volume commercial products.

Upon review of this lead design, the length of the lead from the body to the first lead bond is 0.053 in. This allows for added compliancy and an adequate clearance for lead forming. In this case, the 0.076-in. body height from the substrate surface allows a lid clearance of 0.026 in., which is considered a maximum, to 0.015-in. clearance, which is the minimum. This is based on cleanliness and vibration concerns. That is, a body surface closer than 0.015 in. to the substrate surface would impose cleaning difficulties, and a body surface greater than 0.026 in. would impose vibration and lead damage concerns. The lead's foot length of 0.050 in. provides sufficient length for reflow tooling, lead-to-pad pattern recognition, and a strong lap joint.

Bibliography

Balde, John W., "VLSI Packages: Pin Grids or Chip Carriers," *Circuits Manufacturing,* September 1983.

Brown, R., and Petit, R., "A Miniature Chip Carrier for Microcircuit Applications," *International Microelectronics Symposium,* Philadelphia, Pennsylvania, October 1983.

Burkhart, K. P., et al., "An 18-MHz VLSI Microprocessor," *Hewlett-Packard Journal,* August 1983.

Fennimore, John, "Hermetic Ceramic Chip Carrier Implementation," *Electronic Packaging and Production,* May 1981.

Geschwand, Gary, "A Trend in High Density Packaging of I/O Chip Carriers," *Electronics,* August 1985.

Settle, Roger, "A New Family of Microelectronic Packages for Avionics," *Solid State Technology,* June 1978.

Stafford, John, "Chip Carriers: Application and Future Direction," *Electronic Packaging and Production,* July 1980.

Waltersdorf, Harvey, "Choosing Packages Wisely Pays Off in I/O, Speed, and Space," *Electronic Design,* June 19, 1986.

Winkler, E., "Integrated Circuit Packaging Trends," *Solid State Technology,* June 1982.

Surface Mount Sockets, Connectors, and Component Packaging

4.1 Sockets

Using sockets for any application compromises density, reliability, and costs. They are, however, convenient components for analyzing component and design behaviors for new and untested printed circuit board (PCB) assemblies and for burning in components when required. Ease of component removal, except during prototype stages, and reduction of lead damages are not good reasons for using sockets.

Sockets for SMDs are available with both surface mount and through-hole lead configurations. As shown in Fig. 4.1, there are three popular types of sockets: (1) the AMP through-hole mount socket, (2) the burn-in type surface mount socket, and (3) a true surface mount socket. The AMP through-hole mount socket accepts types A, B, and D leadless chip carriers (LCCs), which are intended solely for socketing. This socket is available in 44, 52, 68, 84, and 100 pins with an overall height of 0.275 in. The snap-on substrate cover is profiled to accommodate heat sinks and provides for access probing when installed.

The burn-in type socket is intended for J-leaded PLCCs and can also be used for permanent mounting. It is available with various pin counts in both through-hole and surface mount lead configurations. A true surface mount socket has J-type leads and is intended for J-leaded PLCCs.

Other more sophisticated sockets exist, such as the AMP 320-I/O surface mount socket intended for a type A or B LCC. This socket has centerline spacings down to 0.010 in. The leads to this socket are not soldered to the board, although they have a flattened out J configuration. Instead, two aluminum plates are bolted to the top of the socket

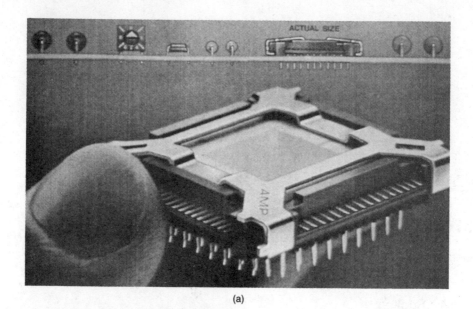

(a)

(b)

Figure 4.1 (a) AMP's through-hole-mounted socket which accepts types A, B, and D LCCs; (b) a popular burn-in type surface mount socket with through-hole leads; (c) a true surface mount socket intended to accept J-lead PLCCs which are soldered to a PCB via its own J leads.

from the opposite side of the printed circuit board; this provides sufficient force to compress the 320 leads against the termination lands on the PCB.

Still another sophisticated socket, developed by Raychem Corporation, is the Cryotact Bubble Memory socket. This socket, as shown in Fig. 4.2, is designed for the most common 1- and 4-Mbit 20-leaded bubble memory packages. It consists of a unique material, a nickel-titanium alloy, which changes shape as it is cooled, allowing for zero insertion forces and absolute continuity during shock and vibration.

4.2   Connectors

Surface mount connectors come in a variety of sizes and configurations. However, the two most important factors regarding the reliabil-

Figure 4.1   (*Continued*)

Figure 4.2   Raychem Corporation's Cryotact zero insertion bubble memory surface mount socket.

ity of surface mount connectors are their mechanical support and their ability to be unaffected by the soldering process.

Mechanical support of the connector's body to the PCB via rivets, heat stakes, nuts and bolts, or other techniques provides mechanical support for the solder joints. Otherwise, when a cable or harness is mated to the connector, continuous stress is imposed on the solder joints, which ultimately causes solder joint cracks as a result of the creeping affects of tin-lead solders.

The material that the connector's body is made of is of great importance since the material must be capable of withstanding infrared radiation (IR), vapor phase soldering temperatures, and dwell times at those elevated temperatures. Liquid crystal polymers, polyetherimide, and polyethersulfone are some of the thermal plastics used for surface mount connectors. Whatever type of material is used, characteristics such as adequate electrical properties, good mobility, dimensional stability, and high glass transition temperatures are the major factors allowing use of a material for a connector body. In any event, performing qualification tests on the surface mount connectors is recommended to determine the effects of soldering and cleaning with solvent.

4.3   Packaging for Surface Mount Components

The most common types of packaging for SMDs are plastic tubes, waffle packages, and tapes and reels. The plastic tubes can be used to package chip resistors, SOICs, PLCCs, and LCCs. They are made of antistatic plastic materials and are the most popular packaging for SOICs and PLCCs.

The waffle packages are plastic conductive or antistatic trays with embossed areas for each component. They are commonly used to carry gull-winged quad packs and currently have not been designed for standardization.

Tapes and reels are used to package all types of SMDs such as chip resistors, chip capacitors, SOTs, inductors, potentiometers, most discrete components, and SOICs. Tape-and-reel packaging has been standardized in specification EIA-481A, which provides tape-and-reel dimensions for 8-, 12-, 16-, 24-, 32-, 44-, and 56-mm size tapes. For high-volume production, this should be the preferred packaging for all types of SMDs with the possible exception of gull-winged quad packs.

Surface Mount Design

Introduction

*Part 2 presents the foundations of surface mount design from
the manufacturability and reliability perspective. These
foundations are based on verifiable research done in the
laboratory and on the manufacturing floor under a wide
variety of conditions and on products ranging from computers
to military assemblies.*

*For most typical commercial assemblies, the information
presented herein may be all the help necessary to develop good
surface mount designs. For complex assemblies, these design
guidelines will also provide sufficient information for the
designer; however, as the printed circuit board (PCB)
construction increases in complexity, the designer should
always seek the advice of an engineer whose expertise lies in
the area of substrate materials and PCB fabrication.
Furthermore, for assemblies that are used in life-threatening,
life-support, or nonretrievable products, the author
recommends that the manufacturability and reliability of the
assemblies be verified with environmental and power cycling
testing which exceeds the actual situation the product will
experience in the field. Also, for these situations, do not
assume that the reliability of the assembly and PCB will be
the same when fabricated and assembled by various
subcontractors or in various factories.*

*Since the 1960s, millions of dollars of research has been
done, primarily by military and aerospace contractors, to
determine the optimal PCB design conditions for SMAs.
These design conditions include various land (pad)
geometries, PCB materials, PCB constructions, and SMD
types. Out of financial and technical necessity, the military
and aerospace pioneers in SMT had to understand how to*

*create optimal conditions with respect to manufacturability
and reliability. The driving force behind this was the fact that
assembly field failures could cause product failures worth
hundreds of millions of dollars or even more catastrophic
results related to the defense of free nations.*

*Based on this initial research, the author has adapted these
design guidelines for high-volume commercial products and
done further research and manufacturing evaluation to
determine the key elements of producing reliable and highly
manufacturable surface mount designs. Because of the vast
array of SMD configurations, it would be naive for anyone to
claim to know the exactly optimal design conditions for every
type of product. However, a reasonably finite set of design
guidelines and concepts has been shown to work well in a
wide variety of situations that are common to all SMAs. In
most instances, a high-volume SMA that can be
manufactured with very high yields can be considered a
reliable one within certain conditions, whereas a "proven
reliable design" that does not help produce high
manufacturing yields will always yield a less reliable or very
unreliable assembly.*

Chapter

5

Surface Mount Design Basics

5.1 The Role of the PCB Designer

Compared to plated through-hole (PTH) technology, the role of the designer in SMT has changed dramatically. What the designer incorporates can drastically affect various areas of the product, ultimately leading directly to its success or failure. The first priority for the designer in SMT is the manufacturability and reliability of the assembly. If the design is suboptimal with respect to manufacturability and reliability, very little can be done in subsequent stages to rectify it. The assembly is doomed to create headaches for the assembler, tester, and ultimate user. The designer must therefore be familiar with the issues that contribute to a good design; these issues, as discussed herein, should predominate when design trade-offs are made. In all cases, a good designer should understand, as much as possible, the manufacturing assembly and PCB fabrication processes and communicate with the manufacturing engineers who will eventually inherit the designer's foundation of work.

The emphasis on a manufacturable and reliable design, simply called an *optimal design*, leads to a basic question that all designers want to know: What are optimal design conditions? Although the answer is not trivial and the question somewhat vague, it is more practical than most people may think. Reading and studying this chapter will provide most of the answers to this question; however, as an outline to the answer, the designer should understand the following major suggestions prior to and during the design phases:

1. The lowest cost assembly will be one which can be fabricated and assembled with zero defects or high yields.

2. Assume that all SMAs must be automatically assembled, which will require specific component-to-component clearances for X, Y, and Z directions.

3. Understand the restrictions imposed by machine automation as it relates to maximum boards or panel size, component-to-board edge clearances, and tooling hole locations and tolerances.

4. When specifying components, solder masks, etc., be sure of their compatibility with the manufacturing process and materials, particularly with the reflow type and solvent cleaning process.

5. When designing land geometries for a specific part number, always review the component's dimensions and tolerances from at least one other source in order to provide a land geometry which will also fit a second source component if necessary.

6. Solder opens and shorts can be the result of too small or too large land geometries, respectively. In most cases, from a manufacturing standpoint, process engineering will play havoc in attempting to correct these problems when design is the cause. Verify all your land geometries and component layouts with manufacturing and process engineering.

7. Make the assembly as easy to test electrically as possible. This does not mean to design in testing such that the assembly must be tested from both sides. Testing from one side is always preferred and least expensive.

5.2 Dispelling SMT Design Myths

Because SMT for commercial products is in its adolescent phase, an abundance of opinions have been expressed about the "correct" design concepts and layout rules. Unfortunately, most of these concepts and rules were developed by committees and by those who have never had to correct design defects on the manufacturing floor. Today, one can go to 12 different designers or design companies and get 12 different land dimensions for a component as simple as a chip resistor, and of course each is backed by a hypothetical reason that justifies why the particular dimensions of one are better than the others. The following suggestions should help the reader to clarify some of the basic issues that are often brought up in the initial design stages:

1. Land geometries and component layouts that work well for vapor phase and IR reflow methods may cause poor results in the dual-wave soldering process. Normally those designs that work well in

dual-wave soldering will also work well in vapor phase and IR reflow processes.

2. Minimizing land sizes to "help" the manufacturer control component movement during reflow is ill-advised. First of all, component movement usually occurs only with low mass components such as 1206-style chips or smaller. Leadless chip carriers (LCCs) also move quite easily, regardless of the size. Second, shrinking land sizes usually carries with it reliability and manufacturability penalties. Third, experienced SMT manufacturers have a variety of process techniques to use for the control of SMD movement during reflow.

3. SMDs do not tend to center themselves on symmetrical pads during reflow. Even if there are some verifiable, repeatable situations where one SMD self-centers, a zero defect process cannot depend on a result that is not predictable.

4. When liquid solder mask is designed in, the artwork should be designed so that the solder mask opening around the conductor or lands is a minimum of 0.010 in. When dry film or liquid photo-imagable solder masks are used, a minimum clearance of 0.005 in. is preferred.

5. Designing in chip SMDs on the bottom side of the substrate for dual-wave soldering is preferred, especially when they are used in large quantities. When compared to vapor phase and IR reflow processes where solder pastes are used, the wave soldering process can provide stronger solder joints and can more easily overcome the solderability problems frequently found with chip SMDs.

6. Substrate thicknesses less than 0.062 in., such as the standard 0.032-in.-thick substrate, will present significant manufacturing difficulties as the square area increases due to substrate warp and twist.

5.3 SMD Selection Guidelines

SMD selection for SMAs is much more complicated than for comparable PTH assemblies, primarily because the variety of SMDs is so large. Furthermore, the types and conditions chosen with a particular SMD can have a dramatic effect on the assembly's reliability and manufacturability. Listed below are some of the most common recommendations that can be used in this area:

1. Use 1206 style packages for resistor and capacitor chips. They are the most common and should present no problem to the manufacturer. Only use the smaller 0805 style when space and electrical performance are absolutely critical. Beware of chips larger than the 1206

style types since cracking of these larger size chips becomes increasingly easier.

2. Avoid using metal electrode face bonding (MELF) packages. These packages look like small glass cylinders with metal end caps and are commonly used for diodes and resistors. Due to a glass-metal seal, they are more susceptible to damage, and due to the round instead of flat curvature of the body, they are more difficult to automatically pick and place. They are commonly used for volume consumer products because they are typically less expensive. When used to package diodes, their electrical characteristics, however, may have some benefits.

3. Avoid surface mount connectors that do not have some means of mechanical support to the substrate surface. Constant strain on the solder joints, which can occur without proper mechanical support of the connector body, can result, over time, in cracked solder joints.

4. Use molded, encapsulated tantalum capacitors instead of the uncapsulated types. The plastic encapsulated types are more reliable and easier to automatically pick and place.

5. Specify low-profile SOT packages which usually have a 0.0005- to 0.004-in. off-contact distance from the substrate. Unlike the moderate- to high-profile SOTs, this type sits more closely on the substrate, which facilitates the use of adhesive bonding if the manufacturer needs to use it.

6. Choose between J-leaded and gull-wing SMDs appropriately. J-leaded SMDs take up less space and have fewer non-coplanar and damaged lead problems during handling and assembly. Gull-wing SMDs provide more compliancy or stress relief and are easier to inspect and rework.

7. Using ceramic LCCs requires mounting these SMDs on substrates that have controlled thermal expansion properties. This is a very costly approach.

8. Require that all SMDs be value-labeled and that the SMD and its labeling be capable of withstanding all forms of soldering and solvent cleaning.

9. Specify solder-coated leaded SMDs. The solderability of solder-coated leads is superior to plated lead finishes.

10. Specify nickel barriers on chip resistors and capacitors. End caps with no barrier metal will normally produce silver-tin intermetallics, which are brittle, leading to joint cracking.

5.4 Types of Surface Mount Designs

The fact that SMDs can be mounted on both sides of the substrate allows several different types of surface mount designs. This gives the designer some added flexibility; however, it also increases the number of decisions the designer must make when starting the design. There are, however, some basic suggestions that the designer should strongly consider.

1. When mixing SMDs and conventional components on the same substrate, try not to locate any type of surface mount quad package (leads on all four sides of the package) on the opposite side from where the conventional components are located. If this occurs, it is likely that the conventional parts will have to be hand-soldered since these types of SMDs currently cannot be wave-soldered without the formation of solder bridges (hot air knife included). It is preferred, but not mandatory, to treat small outline integrated circuit (SOIC) packages in the same way as quad packages; however, it is possible to wave-solder these SMDs.

2. Do not place chip components under plastic leaded chip carriers (PLCCs). If this occurs, visual inspection of the solder joint connecting the chip to the substrate will be impossible. Also, slight variations in the chip thickness may cause the PLCC to raise off the substrate, allowing large gaps between the leads and lands, resulting in weak solder joints or opens.

3. Allow sufficient space between PLCCs (J-leaded types) so that visual inspection and rework are possible. This is normally a 0.120-in. minimum as measured from the outermost dimension of the component.

4. Sufficient space should be allowed between the clenched leads in conventional components and any neighboring SMD.

5. Always allow sufficient clearances from the edge substrate and surrounding components. The clearance will vary depending on the conveyors used to handle the assemblies. In most cases, a panel design is preferred.

Although several more suggestions could be added to the above list, the ones described are typically the ones overlooked or considered unimportant; however, they are very important for manufacturing and quality control.

The next major concern when beginning a design is to choose from the several possible designs the one specific type that best suits the manufacturing process and in some cases the product packaging. When the designer is at this stage, all efforts should be made to choose the design type that will not introduce hand soldering of any type of component to the substrate. Let us now review the major types of surface mount designs.

5.4.1 Design type I

Type I can be classified as the simplest in that SMDs such as chips and
ICs are located on one side of the substrate. In most cases, some conventional components and connectors are present and are located on
the same side as the SMDs. The SMDs on this board would most likely
be attached with solder paste and reflowed by vapor phase or IR methods. The conventional components must always be assembled after the
SMDs have been reflowed and then wave-soldered.

5.4.2 Design type II

Design type II is similar to type I in that the conventional components
and the surface mount ICs are on the same side, but the chip components and sometimes the discrete components (such as SOTs) are on
the solder side of the substrate. This type of design is normally processed by vapor phase or IR reflow soldering the surface mount ICs,
placing the chip and discrete components on the solder side and bonding them with adhesive, and then inserting the conventional components onto the top side. The final wave-soldering process is used to simultaneously connect both the conventional components and the
SMDs on the solder side.

5.4.3 Design type III

Design type III can be considered an assembly that is predominantly
populated by SMDs such as chips, SOTs, PLCCs, and SOICs. In this
case, it is typically impossible for the designer to put all the components on one side; however, efforts should be made to place at least all
the PLCCs on the same side as the connectors or conventional components, if any. This will most likely prevent any hand soldering since if
some of the SOICs are required to be placed on the solder side, it is
still possible to wave-solder these components. If limited real estate
forces the designer to locate the surface mount ICs on both sides of the
substrate, using surface mount connectors would eliminate labor-
intensive hand soldering of conventional connectors, if present. In this
case, either the vapor phase or IR reflow process would be used to connect all the SMDs.

5.5 Orientation Guidelines for SMDs

Automatic machine placement of SMDs such as chips, SOICs, and
PLCCs requires attention to the orientations of these SMDs during
the layout stage.[1] Optimizing SMD layout will increase manufactur-

ing productivity, inspectability, and reworkability. Furthermore, the ease of testing and solvent cleaning should also be enhanced.

The initial concern in regard to SMD orientation is the clearance around the component. Figure 5.1 illustrates the preferred clearances around several of the more commonly used SMDs when they are either vapor phase or IR reflow soldered to the substrate. Figure 5.1a illustrates the clearances for SOICs where a 0.035-in. minimum clearance on the nonleaded side is preferred. This will aid placement and alignment tolerances and prevent interference fits due to extraneous epoxy material which sometimes extends from the component's body. On the leaded sides of the SOIC, a minimum clearance of 0.020 in. from the end of the land is preferred; this will facilitate the placement of solder mask and conductor routing between these areas and allow for solder paste misalignment if it occurs in this direction. Also, a land-to-edge board clearance minimum of 0.200 in. is normally necessary in order to avoid component interference with wave solder fingers and conveyor belt holding mechanisms.

Figure 5.1b illustrates the clearances necessary for PLCCs with J-leaded configurations. Here it specifies a 0.120-in. minimum clearance from lead to lead. This is critical since, as this clearance decreases, visual inspection and reworkability become increasingly more difficult, to a point where neither one is possible. Of course, testing and solvent cleaning become yet another challenge.

Figure 5.1c illustrates the clearances necessary for chip SMDs. For vapor phase and IR reflow methods, the land clearances between other adjacent lands should be a minimum of 0.010 in. Of course, a larger clearance would be preferred; however, less than this may tend to produce solder shorts and component interference on adjacent lands or components. For dual-wave soldering, this same clearance should be increased to 0.025 in. for preferred conditions; however, a clearance of 0.012 in. is possible without wave-soldering defects if components are well aligned.

Still another important concern with SMD orientation is the layout of these components. The preferred orientation occurs when SMDs of the same type are laid out with one axis and one polarity direction, as shown in Fig. 5.2. This minimizes machine placement movements such as head and table rotations and therefore can significantly decrease the component placement, which is sometimes considered one of the most expensive parts of the manufacturing process. As shown in Fig. 5.3, component layouts with several axis and polarity directions will maximize component placement time and also actually increase the level of difficulty for visually inspecting the solder joints.

When SMDs are wave-soldered, additional layout rules are necessary in order to maximize the ability of the soldering process to reach

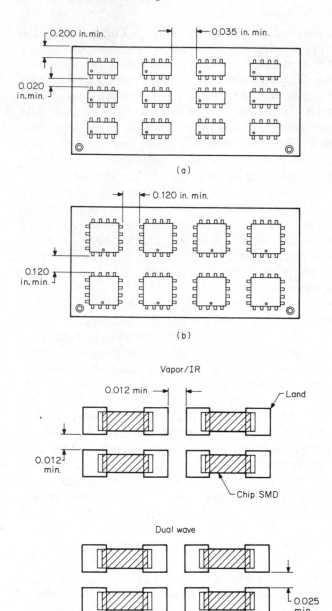

Figure 5.1 Preferred clearances around SMDs when either vapor phase or IR reflow soldered. (a) SOICs; (b) PLCCs; (c) chip SMDs.

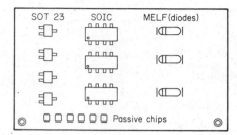

Figure 5.2 Preferred orientation for SMDs—one axis and one
polarity direction.

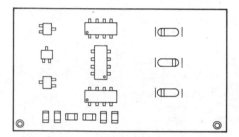

Figure 5.3 Unpreferred orientation for SMDs—several axis and
polarity directions.

zero defects. One of the major concerns relates to the component's lead
or metallization direction to the directional flow of the molten solder.
The other concern, one commonly overlooked, is the close proximity of
tall components such as tantalum capacitors to low-profile components. Figure 5.4 illustrates these additional layout rules, which
should always be incorporated into the design if zero defect manufacturing is one's goal. In both cases, these layout rules will help overcome the formation of solder shorts and opens caused by process limitations and air pockets on the trailing side of the SMD.

5.6 PCB Conductor Routing Guidelines

There are several areas of concern regarding conductor routing on a
surface mount PCB that will have an effect on manufacturability and
reliability. Routing conductors between plated through holes and
lands will be the first major consideration.[1] The low-density design
guidelines for routing a conductor between plated through holes on
0.100-in. centers are illustrated in Fig. 5.5. Here, a 0.015-in.-wide conductor can easily be routed between two 0.060-in.-diameter lands with
0.0125-in. spaces on either side of the conductor. Moderate density, or
two-track, circuit boards fall under this category as illustrated in Fig.
5.6, where two 0.010-in.-wide conductors are routed between two
0.050-in.-diameter lands with reduced spacings of 0.010 in. High density, or three-track, circuit boards, as illustrated in Fig. 5.7, will allow
a total of three conductors between the PTHs with 0.006-in. widths

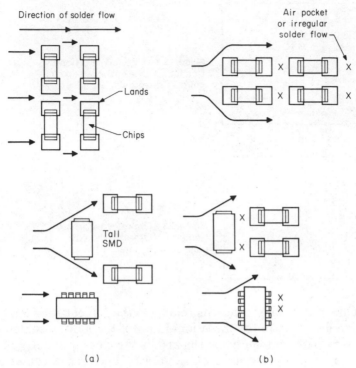

Figure 5.4 Special SMD layout rules applicable for assemblies to be dual-wave-soldered. (a) Good layout; (b) poor layout.

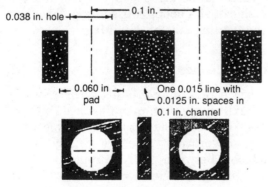

Figure 5.5 One-track, low-density routing dimensions.

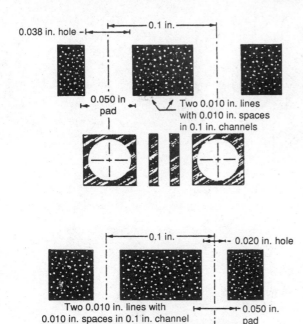

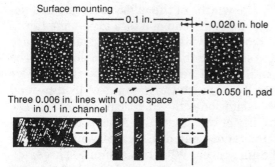

Figure 5.6 Two-track, moderate-density routing dimensions.

Figure 5.7 Three-track, high-density routing dimensions.

and 0.008-in. spaces. In even more complex or higher density designs where adjacent PTHs are on a grid less than 0.100 in. wide, the design parameters for PTH lands and conductor routing can be calculated by using the formula illustrated in Fig. 5.8.

Routing conductors between surface mount lands is another major

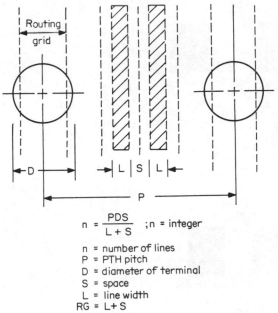

$$n = \frac{PDS}{L+S} \quad ; n = integer$$

n = number of lines
P = PTH pitch
D = diameter of terminal
S = space
L = line width
RG = L + S

Figure 5.8 Formula for routing conductors between plated through-holes.

concern when designing a PCB. When the designer is faced with this condition, it is always preferable to avoid routing conductors between land patterns such as for SOICs and PLCCs. When this occurs, the designer must reduce or modify the widths of the adjacent lands and design in a photoimagable solder mask instead of a less expensive liquid solder mask. The reduced or modified widths of the adjacent lands are necessary to provide adequate space for the conductor. In doing so, the reduced or modified land widths will reduce the amount of solder forming the joint and potentially increase the amount of lead-to-land that is misaligned. These conditions in turn can decrease the manufacturability and reliability of the assembly. Also, routing a conductor through surface mount lands that are normally on 0.050-in. centers will allow only 0.008-in. spaces. With these small spaces, applying solder mask over the conductor is highly recommended in order to prevent solder shorts. However, the application of screen printing liquid solder mask will not provide enough accuracy to prevent the flow of solder mask onto the adjacent lands. This, in turn, will cause significant loss in first-time soldering yields, which is normally why dry film or liquid photoimagable solder masks are used in place of nonphotoimagable solder masks. To reduce or prevent routing between lands and thereby avoid increasing the cost of the PCB or the need for the more expensive photoimagable solder masks, the designer

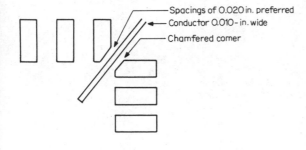

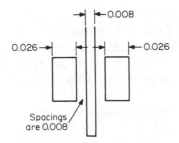

Figure 5.9   Various ways of routing conductors between lands.

should consider finding additional routing possibilities on the surface of the PCB or even routing these particular conductors on other layers of the PCB. In the event that there are no alternatives, then routing the conductor between adjacent lands as shown in Fig. 5.9 is recommended. Keep in mind that only those land widths which are directly adjacent to the conductor should be reduced or modified. All other lands should conform to the normal dimensions as further discussed herein.

Another conductor routing concern is PTH attachment to lands. In this case, to prevent component movement due to solder flow along the conductor and potentially down into the PTH, conductor widths should preferably be between 0.008 and 0.012 in. long at a minimum. In addition, the conductor should be covered with solder mask along with the PTH if it is not used as a test point. Figure 5.10 illustrates the acceptable and unacceptable design conditions related to connecting PTHs to lands.

5.7   Solder Masks and Their Design Guidelines

Solder mask is used to prevent solder bridging between adjacent conductors and to protect the circuitry from mechanical damage and con-

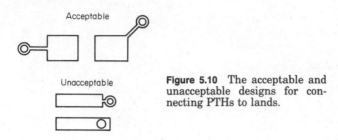

Figure 5.10 The acceptable and unacceptable designs for connecting PTHs to lands.

tamination. In SMT, it is also used to prevent the flow of solder down conductor paths and to cover PTHs so that wave-soldering fluxes are prevented from traveling to the top side of the substrate. It is also used in helping to hold a better vacuum during screen printing and testing. Although the incorrect usage of solder mask can create a variety of manufacturing defects, the costliest errors occur when solder mask is deposited on a land where a SMD will be soldered and when the incorrect type of solder mask is used, depending on the type of manufacturing process.

In general for SMT designs, using liquid solder mask is only appropriate in low- to moderate-density designs where routing traces between pads are uncommon. This limitation is primarily due to the inherent misregistration accuracy of the screen printing process (usually limited to 0.008 to 0.010 in.) and the flow characteristics of the material upon application and curing. Figure 5.11 shows SMD land patterns where screen-printed solder mask flowed onto these ar-

Figure 5.11 Surface mount land patterns that are grossly covered with liquid solder mask and also difficult to detect.

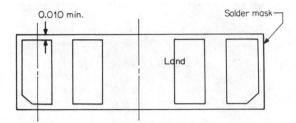

Figure 5.12 Gang-type window design for liquid solder mask.

eas. This PCB defect, which is extremely difficult to see, will undoubtedly produce a solder open between the land and the component lead. Because of the difficulty of visually finding it, this type of problem can easily find its way to the assembly manufacturing level, resulting in costly rework procedures and a reduction in total assembly reliability. To avoid this problem the "gang" solder mask window design, as illustrated in Fig. 5.12, can afford effective board masking and at the same time minimize the chances for liquid solder mask to flow onto the lands. This approach omits placing the solder mask between the lands, which is not necessary since conductors are not routed in these areas. It is not a good or valid reason to place any type of solder mask between lands (such as for SOIC, PLCCs, etc.) when there are no conductors routed between them.

Since liquid solder mask will flow onto lands when designs have conductors routed between lands, dry film or photoimagable liquid solder mask must be used in these cases. The photoimagable liquid solder masks are becoming more popular since they are resilient at high temperatures, encapsulate conductors, and adhere well over epoxy glass and tin-lead plated conductors, unlike dry films. Figure 5.13 illustrates the high-resolution, no-flow properties of photoimagable solder masks when incorporated in high-density circuitry. Furthermore, when using dry film solder masks, the designer should always specify bare copper PCBs, especially when the assembly will be wave-soldered. Figure 5.14 illustrates the poor adhesion of dry film over reflowed tin-lead plated conductors, resulting in the blistering and peeling away of the solder mask. To some, this may look harmless; however, flux entrapment underneath the solder mask is inevitable and potentially results in electrical failures.

No matter which type of photoimagable solder mask is used, the "pocket" solder mask design should be incorporated when conductors are routed between lands. The dimensions associated with this design are given in Fig. 5.15. Using this design will prevent solder bridging to the conductor during any soldering method.

(a)

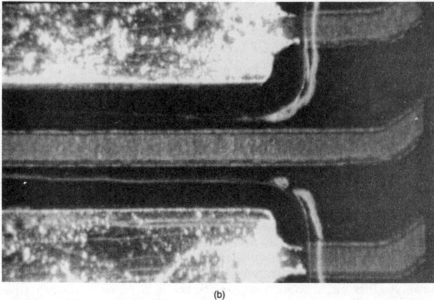

(b)

Figure 5.13   Photoimagable solder mask with high-resolution and no-flow properties. (a) 10 ×; (b) 35 ×.

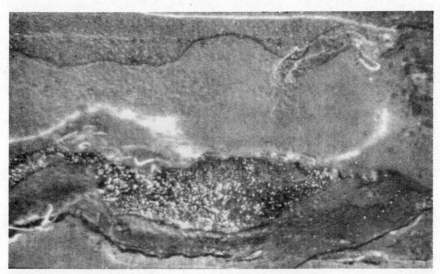

Figure 5.14 The blistering and peeling away of dry film solder mask over a reflowed tin-lead conductor.

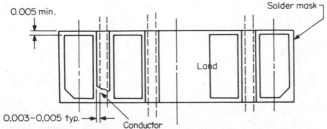

Figure 5.15 Pocket-type window design for photoimagable solder mask.

A last consideration for the design as it relates to solder masks is actually not using them at all. This is common practice in most military and aerospace PCBs because of their higher reliability. To avoid using solder masks, additional conductive layers are produced in the PCB so that the outside layers have only the component land patterns and short conductor runs from the lands to PTHs as necessary. In these cases, all of the long conductor runs are laminated internally, thereby avoiding the need for solder mask on the outside layers. Designing the outside layers in this way also has the additional advantage of allowing more real estate for more SMDs or obtaining the optimum land geometries which otherwise might not have been possible.

References

1. Carmen Capillo, "How to Design Reliability into Surface Mount Assemblies," *Electronic Packaging and Production*, July 1985.

Bibliography

ANSI/IPC-SM-840, "Qualification and Performance of Permanent Polymer Coating (Solder Mask) for Printed Circuits."

Capillo, C., "The Assembly of Leadless Chip Carriers to Printed Wiring Boards," *Nepcon Proceedings,* Cahners Exposition Group, March 1983.

Capillo, C., and Koenig, G., "Designing for SMT," *Printed Circuit Design,* December 1986.

Denkler, John, "Common Questions about Solder Masking," *Electronics,* September 1986.

Fraula, Dennis, "Comparison of Current Solder Masking Technologies," *Electronics,* October 1986.

Losert, Ewald H., "Progress in the Field of Liquid Solder Resists," European Institute of Printed Circuits Summer Conference, 1983.

6

Printed Circuit Board SMT
Design Guidelines

6.1 PCB Construction Guidelines

When leaded SMDs are used, the leads act as the compliant member allowing expansion differences in the component and substrate. These expansion differences are caused by variations in the material's coefficient of thermal expansion (CTE) and in the temperature due to power dissipation within the component. When leadless SMDs are soldered directly to the substrate surface, the solder joint becomes the only available compliant member and therefore receives the stresses resulting from the differential movement of the component's body and substrate material. Since solder has very low mechanical properties, this situation is undesirable; over a period of time, the solder joint may see a sufficient amount of stress to result in a solder joint crack. This type of cracking, which is induced by simple power cycling of the assembly, is illustrated in Fig. 6.1. This kind of design oversight would probably not be discovered until sufficient field usage of the assembly weakened the solder joints to the point where cracks form, leading to failure.

When assembled onto noncompliant substrates such as epoxy fiberglass not all leadless components do have this problem. In fact, stresses due to CTE differences are small if the component's surface area is small. Because of this, small leadless SMDs, such as chip resistors and capacitors, mounted on epoxy substrate do not result in stresses high enough for solder joints to fail. This is why, for most commercial assemblies, the standard epoxy substrates are perfectly acceptable. However, when ceramic LCCs and possibly some very rigid leaded SMDs are used, the designer must determine the various

Figure 6.1 Insufficient joint compliancy during power cycling causing joint cracking between the ceramic LCC and epoxy substrate.

materials and PCB construction that will provide sufficient compliancy.[1]

6.1.1   Substrate materials

Ceramic LCCs have been popular since the early 1970s for use in military and aerospace hardware. Their relatively smaller size as compared to their lead component equivalents provides shorter conductor lengths and higher packaging density, which is considered vitally important to these highly technical applications. Ceramic LCCs, which are normally made of beryllium oxide (BeO), have a CTE of 5 to 7 ppm/°C, whereas epoxy fiberglass laminates have a CTE ranging from 12 to 16 ppm/°C. Due to this large difference between the CTEs of these materials, solder joint cracking would easily result under thermal or power cycling conditions. Therefore, to overcome this reliability concern when mounting LCCs onto epoxy fiberglass laminates, suitable substrate replacement materials and constructions have been developed and extensively evaluated.

From a designer's standpoint the easiest replacement of epoxy fiberglass laminate would be simply another type of laminate, preferably with the same or similar fabrication capability. Figure 6.2 provides a list of various substrate materials and their physical characteristics. Of the substrate types listed, the more commonly used are polyimide fiberglass, epoxy Kevlar, and polyimide Kevlar.

Polyimide fiberglass laminate has been a popular substrate material in military and aerospace electronics because it has a higher glass

Figure 6.2   Various Substrate Materials and Their Physical Characteristics

Substrate material	Glass transition temperature T_g, °C	Coefficient of thermal expansion (CTE), ppm/°C	Lateral thermal conductivity, Btu/h · ft · °F	XY tensile modules, psi × 10^{-6}	Moisture absorption, % by weight
G-10 epoxy fiber-glass	125	14–18	0.2	2.5	0.10
G-30 polyimide fiberglass	250	12–16	0.2	2.8	0.35
Epoxy Kevlar	125	6–8	0.13	4.4	0.85
Polyimide Kevlar	250	5–8	—	4.0	1.50
99.5% BeO	—	5–7	120	44.0	—

transition temperature (T_g) than the epoxy fiberglass laminates. This thermal characteristic provides an added advantage during solder joint rework since the peel strength of the laminated copper is higher at elevated temperatures and the Z-axis expansion of the substrate is lower at elevated temperatures, resulting in less stress to the PTHs. Despite these advantages, polyimide fiberglass laminates still do not always provide a sufficient reduction in the CTE, the preferred goal when mounting ceramic LCCs.

The other laminate alternatives consist of either epoxy Kevlar or polyimide Kevlar. Due to the negative CTE properties of Kevlar (−2 to −4 ppm/°C), these laminates provide a sufficient reduction in CTE close to that of BeO, resulting in a built-in laminate compliancy. The only major drawbacks with the Kevlar base laminates are their cost, which is not normally a major concern with military or aerospace electronics, and their recognized fault of internal microcracking. This internal microcracking occurs as a result of the large CTE mismatch between the resin (epoxy or polyimide) and the Kevlar fibers. However, various recognized experts now state that the new epoxy Kevlar laminates provided by General Electric and TRW have overcome this microcracking problem.[2,3] On the other hand, laminates with microcracking have been used throughout the industry without any reported detrimental affects.

6.1.2   Substrate CTE restraining cores

Although the epoxy and polyimide Kevlar laminates can resolve the CTE mismatch for assemblies populated with ceramic LCCs, other reliability concerns such as excessive moisture absorption and micro-cracking have brought about further investigations.[2-4] These include using a low CTE restraining core consisting of either epoxy or polyimide fiberglass within the multilayer board. These CTE restraining cores or foils can consist of several types of materials as illustrated in Fig. 6.3. When laminated close to the surface of the substrate, their low CTE actually restrains or reduces the CTE of the laminate just above it. Therefore, by bonding such a material within an epoxy or polyimide laminate, the high CTE of these materials can be reduced or tailored to a point where they can match or come close to the CTE of the ceramic LCCs. This in turn obtains the goal of compliancy and thus higher reliability.

The most commonly used CTE restraining core consists of various ratios of copper-Invar-copper (CIC). CIC consists of outer layers of copper metallurgically bonded to a core of ultra-low-CTE Invar. Invar is 36 percent nickel, 64 percent iron alloy that is derived from the same family of alloys as Kovar, and alloy 42. Bonding is accomplished by the same process used to make thermostat metals and coinage. By varying the thickness ratios of the high CTE copper to the low CTE

Figure 6.3 Various Types of CTE Restraining Cores and Some Physical Characteristics

Restraining core material	Coefficient of thermal expansion (CTE), ppm/°C	Modules of elasticity, psi × 10^6	Lateral thermal conductivity, Btu/h · ft · °F
Copper-Invar-copper (20%/60%/20%)	6.3	20	95
Copper-Invar-copper (16%/68%/16%)	5.0	19.5	81
Alloy 42	5.3	21	8.8
Copper-moly-copper (13%/74%/13%)	6.5	39	120
Low carbon steel	10	30	27
Graphite fiber, P-75	− 0.75 (ppm/°F)	75	90
Graphite fiber, P-100	− 0.9 (ppm/°F)	105	300
CDA 101 copper*	17.3	17	226
6061 aluminum	23.6	10	110

*Not used as restraining cores.

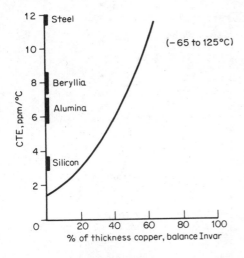

Figure 6.4 Coefficient of thermal expansion (CTE) versus copper content in CIC.

Invar, the core's CTE can be tailored to range from that of silicon to that of beryllium, as illustrated in Fig. 6.4.

Texas Instruments has developed the manufacturing capability to fabricate CIC cores in widths up to 25 in. and thicknesses down to 0.006 in. This permits the use of conventional, high-volume lamination presses and procedures in the manufacture of multilayer boards. The common implementation of CIC within a multilayer board construction normally consists of two 0.006-in. CIC cores laminated near the surfaces of the multilayer board. This type of construction is illustrated in Fig. 6.5, which also uses the CIC as power and ground planes. This approach, however, causes another problem which has presented itself when CIC cores are plated onto the barrels of PTHs. Unless special fabrication processes are used, voids and excessive etchback can occur at the barrel-CIC interface. Figure 6.6 illustrates both of these problems associated with connecting CIC to the PTHs for the additional use as power and ground planes.

Electron-dispersive x-ray spectroscopy of the voids in Fig. 6.6 show that iron, nickel, and copper are not present. Vacuum techniques revealed that the voids are empty spaces without solids; that is, liquid resin penetrated the void when the sample was under a vacuum. Cross sections of the PTHs in the horizontal direction show that voids are caused by the lack of adhesion between the copper plating and the Invar. During acid cleaning it is possible for the iron to react with the acid, yielding hydrogen gas and iron compounds, thereby inhibiting the copper deposition process. Therefore, complete removal and neutralization of these acids is necessary prior to electrolysis copper deposition. This solution may be more simply said than done since various fabricators have not yet demonstrated their ability to prevent

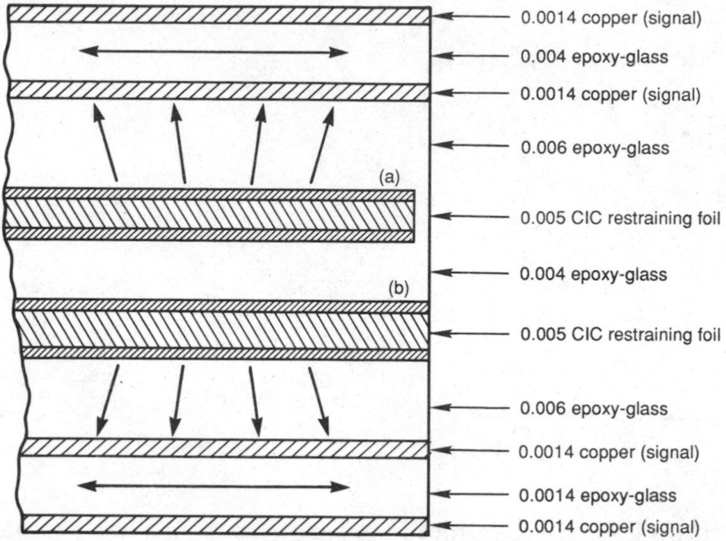
0.0014 copper (signal)
0.004 epoxy-glass
0.0014 copper (signal)
0.006 epoxy-glass
(a)
0.005 CIC restraining foil
0.004 epoxy-glass
(b)
0.005 CIC restraining foil
0.006 epoxy-glass
0.0014 copper (signal)
0.0014 epoxy-glass
0.0014 copper (signal)

Figure 6.5 Typical multilayer board design with CIC restraining foils. (*a*) Recessed from the barrel when not used as power and ground planes; (*b*) connected to the barrel when used as power and ground planes.

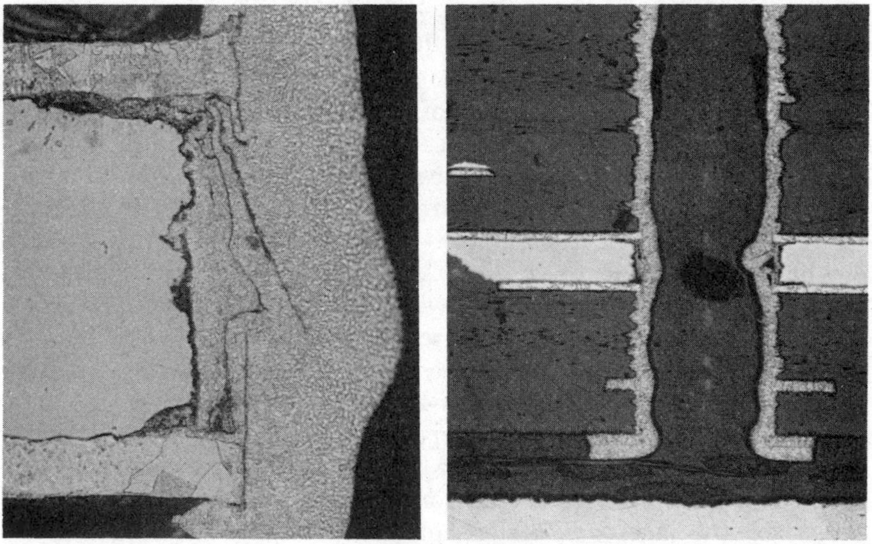

Figure 6.6 CIC core interface at PTH barrel with plated voids and loss of copper adhesion.

these defects from occurring. To avoid this potential reliability problem when plating copper at the CIC-barrel interface, it is strongly recommended that the CIC cores be predrilled larger than the PTHs and thereby designed not to electrically connect to the barrels.

Other restraining cores such as copper-molybdenum-copper (CMC) and graphite are sometimes used. In some cases where high power assemblies require high levels of thermal transfer via conduction through the restraining core, CMC is used in place of CIC. CMC has a much higher thermal conductivity than CIC; in fact, Invar is a very poor conductor of heat. When used as a restraining core, graphite has demonstrated some interesting qualities which neither CIC nor CMC have.[2] Graphite's additional physical characteristics, such as its high thermal conductivity, its high modulus of elasticity, and especially its light weight, make this material ideally suited for spacecraft. Graphite's much higher cost than CIC or CMC has restricted its use in highly technological applications. Also, loss in laminate adhesion to the graphite core has further complicated the use of this material.

6.1.3 Design for thermal transfer

Due to higher packaging densities, miniaturization, and faster circuit speeds, thermal transfer on SMAs is another challenge the designer may be faced with. Heat is generated within the SMD at the die in various amounts depending on the amount of wattage generated. When heat is generated at the die, most of it is conducted through the materials that present the path of least resistance. Figure 6.7 illustrates the possible pathways heat may take through a VLSI component. The most likely pathways or areas to conduct the most heat are through the die attach adhesive under the component and through the ceramic body underneath the adhesive. Depending on how the component is designed, a pathway most likely to conduct most of the heat is the lead frame and the leads extending from the component's body.

Using this simple version of heat conduction through a SMD, one can now design various methods to help absorb and disperse heat from an operating SMD so that it runs below its recommended maximum

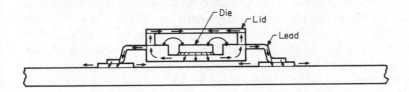

Figure 6.7 Possible pathways heat conducts itself through a VLSI component.

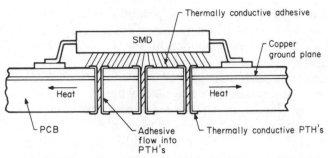

Figure 6.8 Thermally conductive PTHs and adhesive used to dissipate heat from the SMD into a ground plane.

operating temperature. One of the simplest methods, and possibly the least expensive, is to design nonelectrical thermally conductive PTHs in various patterns just underneath the SMD. As illustrated in Fig. 6.8, several thermally conductive PTHs which should also be connected to a power or ground plane are within close proximity to the area of the component generating the heat. Since air is a poor conductor of heat, the PTHs must be mechanically connected in some way to the bottom side of the SMD. This is accomplished by placing thermally conductive adhesive underneath the SMD prior to its placement on the substrate. There are two concerns related to this heat transfer technique. First, the adhesive used must have a curing schedule that does not affect the behavior of the solder paste that will be on the lands during the time the adhesive is cured. A possible approach for an adhesive with a curing schedule (temperature and dwell time) slightly longer than what is required for the solder paste is to cure the adhesive in a two-stage process. The initial stage would consist of partial adhesive curing during the same time the solder paste is being baked out. The final adhesive curing stage would occur during the vapor phase or IR reflow process. The second related concern in using thermally conductive adhesives underneath SMDs is the potential of losing compliancy. To avoid this, flexible thermally conductive adhesives are incorporated.

Another, more common heat transfer method is to use metal heat sinks bonded either on top of or underneath the SMD. Those that are bonded to the top of the SMD are available in various shapes and sizes depending on the size of the SMD and the amount of heat being dissipated. Most types of heat sinks designed to mount on the top of the SMD have fins or grooves which increase the surface area so that additional heat is released to the air circulating within the system. The heat sinks designed to mount underneath the SMD are typically custom-made and are normally used only when the area above the

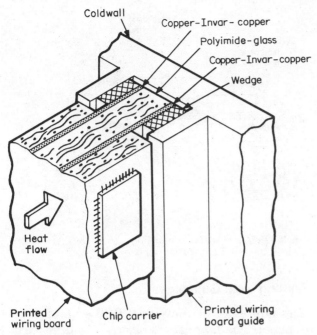

Figure 6.9 Transferring heat via a laminated metal re-straining core to a cold wall.[3]

component is restricted. With this type of design, the heat from the SMD travels through the bonding adhesive and dissipates through the heat sink and down the heat sink leads that are soldered onto thermally conductive PTHs.

A more complicated and expensive method for conducting heat from SMDs is laminating either thick restraining metal cores or corrugated fin stock within the multilayer PCB. The metal cores, which usually consist of 0.062- or 0.042-in.-thick CIC or CMC, are laminated between two multilayer PCBs electrically connected to each other by an edge wrapped around the surface mount connector. As shown in Fig. 6.9, the heat flow is designed to travel down from the SMD, through the PCB, and into the metal core. The metal core is always designed so that it protrudes beyond the PCBs and is then mechanically connected to a cold wall or chassis. The major drawback with this type of thermal design is the excessive weight of the thick metal core. When this factor is a problem, such as for airborne or space flight hardware, corrugated fin stock is used as a replacement. As shown in Fig. 6.10, corrugated fin stock is used in the same way as metal core with the exception that circulating air through the corrugated fin stock is the method of heat removal instead of conduction at a cold wall.

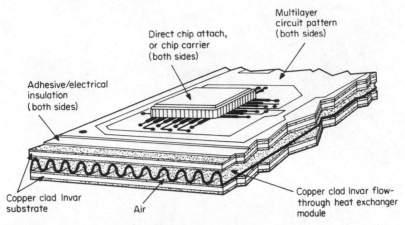

Figure 6.10 Corrugated fin stock laminated within a multilayer board for thermal transfer via circulating air. (*Courtesy Texas Instruments.*)

6.2   PCB Panel and Tooling Hole Guidelines

Most PCB manufacturers fabricate the individual PCB on a 12 × 18, 18 × 24, or 20 × 24 in. panel. The fabricator's objective is to place as many individual PCBs as possible on one panel. The usable area on an 18 × 24 in. panel is 17 × 23 in., because a 1-in. border is required for processing. If the PCB design is 10 × 10 in., the panel yield is two PCBs. If the PCB design is 8 × 10 in., the yield would be four boards per panel. So for basically the same cost, one could now have four PCBs instead of two. This obviously is another area where the designer can actually start off with cost reductions. That is why the designer and packaging engineer should discuss exactly what will be required for board size, understanding that a small dimensional change may have a significant effect on PCB price.

Today, surface mount PCBs are usually small and sometimes irregularly shaped. From an assembly manufacturing standpoint, this imposes additional handling and sometimes fixtures so that the single PCB can be effectively supported during processing. The designer must therefore always consider the total PCB assembly panel design instead of a single PCB design, whether for low- or high-volume manufacturing. An assembly panel design that incorporates several PCBs should usually not exceed 13 × 13 in. This panel design will fit most SMD placement systems and should still have sufficient rigidity. When two PCBs are too big to fit on a panel or when the PCB has an irregular shape, another approach is to design in "breakaway tabs" to aid in the handling during assembly manufacturing. Figures 6.11 and 6.12 illustrate several examples of designing PCB panels and breakaway tabs. Close attention should be paid to the actual dimensions associated with each approach.

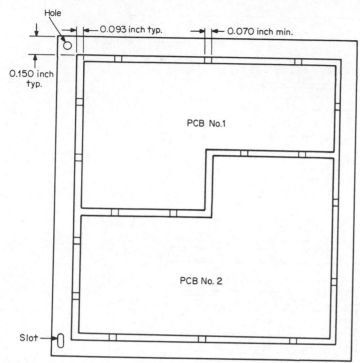

Figure 6.11 A panel consisting of two PCBs enclosed in a standard boarder frame.

Tooling hole designs can be a factor in preventing the corrective and accurate use of automated equipment used to assemble and process SMAs. The standard tooling hole diameter is 0.125 in.; however, most foreign component placement machines require hole diameters smaller than this. In any case, the tooling holes should never be plated and must have a tolerance of −0.000, +0.003 in. for best results. These tolerances should not be a problem from reputable PCB fabricators; however, as PCB tolerances decrease, PCB prices increase. Furthermore, another important design related to tooling holes is that one hole should be circular and the other slotted. The slotted hole should be designed so that the smallest dimension is equal to the diameter of the circular hole and the longest dimension a minimum of 0.020 in. greater than the smaller dimension. This will allow for tolerance buildup over the distance between the hole and slot. If two circular holes are used, boards that are slightly out of dimension may be completely unusable on the automated machinery unless redrilling the holes is performed. If this condition occurs, it could by catastrophic because of the time required to redrill large quantities of boards.

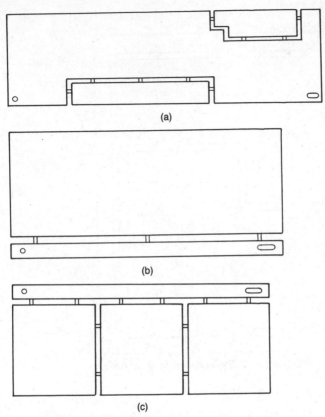

(a)

(b)

(c)

Figure 6.12   Several types of "breakaway tabs." (a) To overcome irregularly shaped PCBs; (b) to provide an area for tooling holes; (c) to provide added rigidity for several attached PCBs.

6.3   Land Geometries for SMDs

Few aspects of surface mount design are more important than land geometries. Optimal land design takes into consideration tolerances in artwork, SMD dimensions, automatic placement systems, and type of reflow process. The sum of good designs is high manufacturability and reliability, which of course are two major cost factors.

Currently, there are several commercially available surface mount design specifications.[5,6] Any one of these would provide the novice with sufficient background in the subject. However, most of these design specifications, all of which have been reviewed by the author, would require revision to achieve zero defect manufacturing (from a design standpoint only) and, in some instances, adequate levels of reliability.

The following sections provide recommended land geometries for various types of SMDs. In all cases, these land geometries have been used in millions of assemblies installed in computers, medical products, VLSI test equipment, aircraft, and spacecraft. Furthermore, they are designs that have been modified and remodified since the 1970s under various manufacturing processes and experiences.

6.3.1 Land geometries: chip resistors and capacitors

The most important land dimension for passive chip SMDs is the extension past the end cap, allowing for proper formation of a reliable solder fillet. The distance between the lands is not as critical but should strike a balance between allowing for placement inaccuracies and providing adequate space for adhesive application.

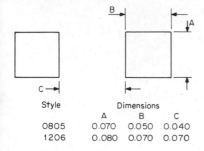

Style	Dimensions		
	A	B	C
0805	0.070	0.050	0.040
1206	0.080	0.070	0.070

For larger or less commonly sized chips, the land geometries can be calculated from the following illustration:

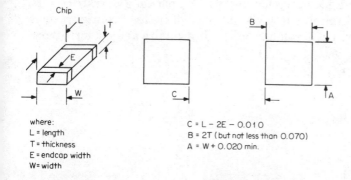

where:
L = length
T = thickness
E = endcap width
W = width

C = L − 2E − 0.010
B = 2T (but not less than 0.070)
A = W + 0.020 min.

6.3.2 Land geometries: tubular SMDs or MELFs

Land geometries for tubular SMDs should be calculated by using the illustration for larger or less commonly sized chips. The author strongly recommends against using the design specified in IPC-SM-782. This design uses a U-shaped land, supposedly to "cradle" the tubular SMD. This idea does not work and reduces the reliability since the bottom side of the end caps are supposed to rest directly on the solder mask or selective material.

6.3.3 Land geometries: SOT 23s

For vapor phase or IR reflow methods, use the following:

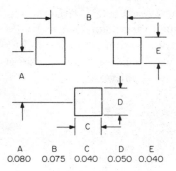

A	B	C	D	E
0.080	0.075	0.040	0.050	0.040

For dual-wave, vapor phase, or IR reflow methods, use the following:

A	B	C	D	E
0.080	0.075	0.050	0.050 – 0.060	0.050

6.3.4 Land geometries: SOT 89s

The following land geometry for the SOT 89 package is restricted for vapor phase or IR reflow methods. Because of its lead configurations, this component will not solder well with a dual-wave soldering process.

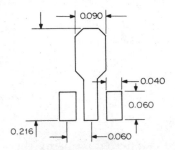

6.3.5 Land geometries: SOT 143

For vapor phase or IR reflow methods, use the following:

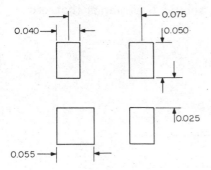

For dual-wave, vapor phase, or IR reflow methods, use the following:

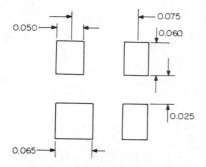

6.3.6   Land geometries: SOICs

Because of the wide variety of SOIC dimensions, especially between the lead heels and the length of the lead's foot, the following formula should be used in calculating the land geometries:

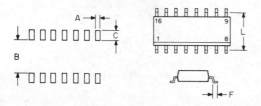

A = 0.030 (0.025 when conductors are routed
    in-between or when wave soldered)

B = L - 0.040

C = 0.075 typ. or F + 0.040

SOICs that are dual-wave-soldered should have lands that are a maximum of 0.026 in. wide.

6.3.7 Land geometries: PLCCs (J-leaded)

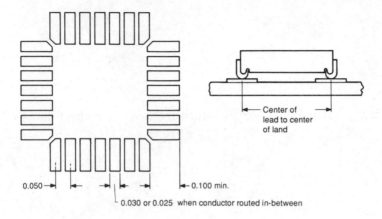

Center of lead to center of land

0.050   →| |←   →| |← |← 0.100 min.

└ 0.030 or 0.025 when conductor routed in-between

6.3.8 Land geometries: leadless chip carriers (LCCs)

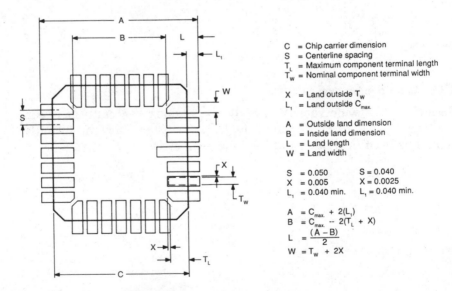

C = Chip carrier dimension
S = Centerline spacing
T_L = Maximum component terminal length
T_W = Nominal component terminal width

X = Land outside T_W
L_1 = Land outside $C_{max.}$

A = Outside land dimension
B = Inside land dimension
L = Land length
W = Land width

S = 0.050   S = 0.040
X = 0.005   X = 0.0025
L_1 = 0.040 min. L_1 = 0.040 min.

$A = C_{max.} + 2(L_1)$
$B = C_{max.} - 2(T_L + X)$
$L = \dfrac{(A - B)}{2}$
$W = T_W + 2X$

6.4 Design for Testability

With conventional assemblies, an in-circuit test is performed by probing the component leads which are on 0.100-in. centers. With SMT, designing in tests becomes a much greater challenge since there are a variety of SMDs, both leaded and leadless, most of which have leads or metallizations on 0.050-in. centers or less.

The first rule to learn is to try to avoid probing directly on component leads or metallizations. Probes that normally apply 4 to 8 ounces of pressure can actually push an open lead down against its land and create a false reading. Ceramic components are very fragile, so probing directly on one of these could possibly crack or damage its body, which might not show up immediately during testing. Furthermore, probing the solder joint should also be avoided whenever possible. Probes that make contact with the solder joint probably will not damage it, but, due to the small size and steep angles of the solder joints, the probe could slide or snap off and possibly short two points together. Damage to the probe can take place over time due to excessive side loads if the probe slides down a solder fillet. This in itself can cause a costly nightmare that would likely require a redesign of the PCB and test fixture. The best remedy for these potential test problems is to design in a remote test land (pad) or an extended land that is attached to the solder joint. Figure 6.13 illustrates these recommendations.

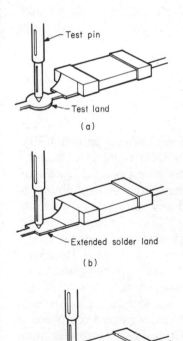

Test pin

Test land

(a)

Extended solder land

(b)

(c)

Figure 6.13 Various areas for probing on a SMD during an in-circuit test and their recommended levels of use. (a) Recommended; (b) acceptable; (c) not recommended.

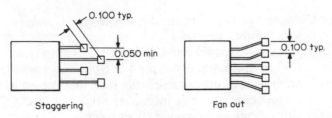

Figure 6.14 Test for land positioning for surface mount ICs.

The test land where the probe will make contact should have a 0.040-in. minimum diameter and should have solder as a contact area. When vapor phase or IR reflow soldering is used, filling the holes will not be possible; however, the test engineer will have to choose the best probe design so that it will not get stuck as it makes contact with the hole. A single pointed probe that has a larger diameter than the hole opening is usually preferred.

When surface mount ICs are tested, the test lands should be 0.100 in. apart if possible, but in all cases no closer than 0.050 in., as illustrated in Fig. 6.14. Furthermore, the test lands should be placed such that their centerline is 0.060 in. away from the edges of SMDs that are less than 0.200 in. high.

6.5   Design as It Affects Reliability

6.5.1   Land geometries

The land geometry, or size, can have significant effects on reliability, no matter which type of land configuration a SMD may have. The land geometry determines or produces several conditions related to how reliable the solder joint connection will be. The land area contributes to the amount of solder which may be applied and allowed to connect to the lead or metallization (for leadless SMDs). This in turn contributes to the volume, or mass, of solder which makes up the solder joint. Of course, the application of solder to the lands, such as with screen printing of solder pastes and wave soldering, can have the greatest effect on the amount of solder deposited; however, it is always likely that the larger the land, the more volume of solder it is able to hold. This leads to the fact that, in most cases, the solder joints composed of larger volumes of solder tend to be more reliable or stronger than those with lesser volumes of solder. Figures 6.15 and 6.16 illustrate this condition as it relates to gull-wing and leadless SMDs, respectively. As illustrated for gull-wing SMDs, the higher the "heel" of the solder joint, the higher the tensile strength of the solder. There is a

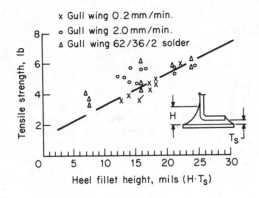

Figure 6.15 Solder heel fillet height as it relates to the tensile strength of the lead pull.[7] (*Courtesy AMP, Inc., Harrisburg, Pennsylvania.*)

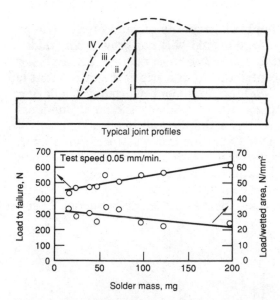

Typical joint profiles

Figure 6.16 Solder mass as it relates to the load to failure for a solder joint and leadless SMD.[8]

limit to this condition in that, as the heel becomes too high, it significantly reduces the lead compliancy, thereby reducing the tensile strength to failure. As illustrated for a leadless SMD, the larger solder joints will have greater strength than the smaller ones. This has been proved for all types of leadless SMDs; however, too much solder will be detrimental because of the chance of shorting and obscuring dewetting or nonwetting soldering conditions to the metallizations or lands. The designer must therefore provide a land area large enough that it does not prevent manufacturing from applying, or make it dif-

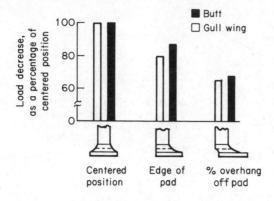

Figure 6.17 Predicted load with various lead positions on land.[7] (*Courtesy AMP, Inc., Harrisburg, Pennsylvania.*)

ficult for manufacturing to apply or vary, the amount of solder to be deposited. Furthermore, too large a land will not provide any added reliability and will only add to a decrease in component density.

The area of the land also contributes to the size of the target that is available for component lead-to-land alignment. Of course, the larger the land, the easier it is to place all the leads on the land. The ideal position for any lead is the center of the land without any portion of the lead overhanging the land. Experimental tests performed by J. Hoyt of AMP show various joint strengths related to the lead position on the land.[7] The results of these tests are illustrated in Figs. 6.17 and 6.18. In all cases, when a lead comes in line with the edge of the land, it loosens the solder joint on that side, resulting in a reduction in sol-

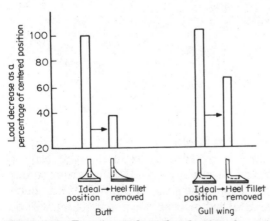

Figure 6.18 Experimental results showing the reduction in tensile strength as a function of lead position on land with lead movement to the back of the land.[7] (*Courtesy AMP, Inc., Harrisburg, Pennsylvania.*)

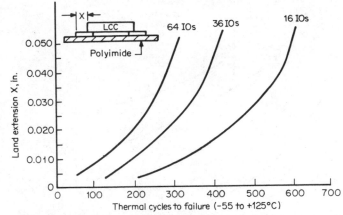

Figure 6.19   Land extension variations as they relate to thermal cycles to failure for ceramic LCCs.[1]

der joint strength. When the lead starts to overhang the land, a further reduction in solder joint strength results.

As with leaded SMDs, the positioning of leadless SMDs on lands affects reliability. With leadless SMDs, the critical land area that affects reliability the most is the amount of land extension beyond the edge of the component. This, of course, is coupled with the amount of solder applied in this area in that more "robust" joints yield higher levels of reliability. Without sufficient land extension in this area, the preferred amounts of solder will be impossible to apply. Figure 6.19 illustrates the change in thermal cycles to failure for LCCs with 64, 36, and 16 I/Os when the land extension varies in length from the edge of the component. Upon review of these data, land extensions less than 0.045 in. produce an increase in the reduction of the reliability, and an increase in land extensions greater than 0.045 in. increases the reliability only slightly, if any.

6.5.2   Substrates

The choice of substrates the designer should consider depends on the type of SMDs to be used and the material they are made of. The major concern, as mentioned, is the degree of CTE mismatch between that of the component and the substrate. Normally, when mounted on epoxy fiberglass substrates, leaded SMDs or any small chip SMD or LCC constructed of organic (epoxy, epoxy polyimide) fiberglass do not impose any significant reduction in reliability.

Typically, the only instances when substrate materials with lower CTE are necessary are when ceramic LCCs or very high power components with stiff leads are used. Figure 6.20 illustrates the change in

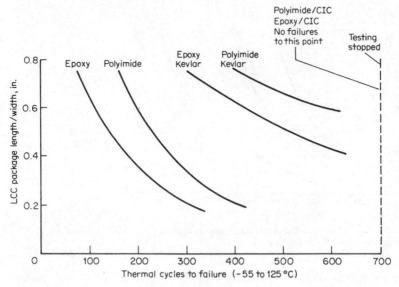

Figure 6.20   The number of thermal cycles (− 5 to 125°C) to failure when ceramic LCCs are soldered to various types of substrate materials and constructions.

thermal cycles to failure when ceramic LCCs are mounted onto various substrate materials. It should also be noted that as the I/O count on the LCC increases, the number of thermal cycles to failure decreases, no matter which substrate is used.

References

1. Carmen Capillo, "How to Design Reliability into Surface Mount Assemblies," *Electronic Packaging and Production,* July 1985.
2. Joesph Leibowitz, William Winters, and Jack Kolkin, "Graphite Layers in SMT Boards Control Thermal Mismatch," *Electronic Packaging and Production,* June 1985.
3. Louis J. Broccia, "Constructing PWBs with Copper-Invar-Copper," *Printed Circuit Fabrication,* July 1986.
4. Cynthia Jones, Rich Yeager, and Foster Gray, "Constraining Core Techniques for Surface Mounted Components, Phase IIIA," *IPC Technical Review,* December 1984.
5. IPC-SM-782, "Surface Mount Land Patterns (Configurations and Design Rules)."
6. Anatrek, "Focus on SMT Design," Santa Barbara, California.
7. J. Hoyt, "Influence of Leg Shape and Solder Joint Metallurgy on Surface Mount Solder Joint Strength," 11th Annual Electronics Manufacturing Seminar, Naval Weapons Center, China Lake, California, February 1987.
8. S. P. Hawkins, C. J. Thwaites, and M. E. Warwick, "The Mechanical Properties of Soldered Joints to Surface Mounted Devices," *Brazing & Soldering,* No. 10, Spring 1986.

Bibliography

Bierman, Howard, "SMDs Tax Board Testers," *Electronics Week,* February 25, 1985.

Bloechle, D. P., and Schoenberg, L. N., "A Low Expansion MLB Structure for Leadless Chip Carrier Applications," *Proceedings of the Technical Conference,* 3rd Annual IEPS, October 24, 1983.

Capillo, C., "The Assembly of Leadless Chip Carriers to Printed Wiring Boards," *Nepcon Proceedings,* Cahners Exposition Group, March 1983.

Capillo, C., and Koenig, G., "Designing for SMT," *Printed Circuit Design,* December 1986.

Dance, F., and Wallance, J., "Clad Metal Circuit Board Substrate for Direct Mounting of Ceramic Chip Carriers," *Proceedings of the 1st Annual Conference,* IEPS, November 1981.

Engelmaier, Werner, "Effects of Power Cycling on Leadless Chip Carrier Mounting Reliability and Technology," *Proceedings of the 2nd Annual Conference,* IEPS, November 1982.

Engelmaier, Werner, "Hi-Rel SMT Solder Joints," *Circuits Manufacturing,* December 1988.

Gray, Foster, "Substrate for Chip Carrier Interconnections," *IPC Workshop Proceedings,* December 10, 1984.

Lassen, C. L., "Use of Metal Core Substrates for Leadless Chip Carrier Interconnection," *Electronic Packaging and Production,* March 1981.

Lau, J., Rice, D., and Avery, P., "Nonlinear Analysis of Surface Mount Solder Joint Fatigue," *Technical Proceedings, IEEE,* 1986.

Sanscrainte, Ren, "Understanding the Manufacturing Process," *Printed Circuit Design,* April 1986.

Smith, William, "The Effect of Lead Coplanarity on PLCC Solder Joint Strength," *Surface Mount Technology,* June 1986.

Thwaites, C. J., "Some Metallurgical Studies Related to the Surface Mounting of Electronic Components," *Circuit World,* Vol. II, No. 1, 1984.

Assembly

Introduction

The manufacturing of surface mount assemblies (SMAs) is a very demanding technological endeavor requiring a sophisticated understanding of the materials, processes, and machinery. The major key to successful manufacturing of SMAs is knowledge; however, there are those who feel that once all the greatest equipment is purchased and set up, success in manufacturing SMAs is only a short distance away. Approaching surface mount manufacturing in this way will be the shortest road to disaster, as some have already found out.

There are three possible approaches in a company's endeavors to start manufacturing SMAs. The first is to develop the technology internally over a period of time. This should involve a team approach where purchasing, design, manufacture, and quality groups all take part in the form of research and development (R & D) projects. When it encompasses the analysis of designs, materials, processes, and machinery, this approach usually requires a minimum of 3 to 4 years and possibly more depending on how it is executed. This probably is the most costly approach.

The second approach is to hire the industry experts with extensive R & D background and hands-on experience in all phases of surface mount manufacturing. This is probably a better way than the first approach since learning curves normally require significant periods of time and can be very costly. In this case, however, the difficult part is determining the best expert. A true expert in SMT, one capable of establishing the technology in a company without it or improving an existing facility, will easily stand out among the so-called experts. The major ingredients consist of a strong background in PCB design and fabrication, metallurgy, assembly reliability, chemistry, and all the processes and machinery, as well as in process controls and their implementations. Experts with real hands-on and development experience (involvement is not experience) will still be few in numbers for many years to come.

The third approach to surface mount manufacturing is simply subcontracting the work. For new and complex technologies, this is typically the best and, initially , the least costly approach. Actually, making use of knowledgeable subcontractors can add value not only as a manufacturing group but also as a technological transfer group. In this case, do not judge a subcontractor's ability by its size but, again, by its depth of knowledge and active process controls within.

No matter which approach a company takes to manufacturing SMAs, a successful pathway will encompass not just machinery but also an experienced team of people dedicated to perfecting all the aspects of design, components, materials, and processes involved with SMT manufacturing. Any one part of this team who does not clearly understand the variables which affect successful manufacturing will undoubtedly cause failure for the entire group.

Part 3 covers various aspects of manufacturing SMAs: its techniques, materials, processes, machinery, and workmanship.

Manufacturing Controls and Incoming Inspection

7.1 Manufacturing Process Sequences

The type of surface mount design being assembled dictates the sequence of steps in the manufacturing process. There are three major types of surface mount designs, as discussed in Chap. 5. Type I, which has SMDs on the top side without conventional components, has the following manufacturing process sequence:

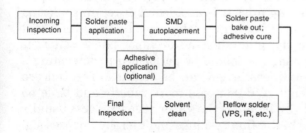

When conventional components coexist with SMDs on the same side, the manufacturing sequence can proceed in two possible ways depending on the type of SMDs. With passive SMDs, PLCCs with J leads, or components which require solder paste application, the manufacturing sequence is as shown at the top of the next page.

This manufacturing sequence shows two possible pathways that can be taken after reflow soldering the SMDs and prior to conventional component insertion. Pathway A incorporates a solvent cleaning step just after reflow soldering. This is preferred when there is a long dwell time or time to finish assembling the conventional components. If there is a long dwell time, then the flux residues left on the assembly

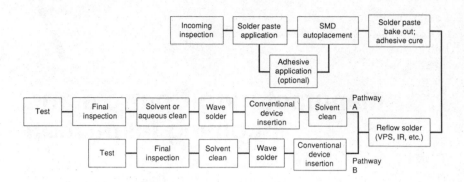

during reflow solder will be extremely difficult to remove effectively during a final cleaning step after wave soldering the conventional components. Furthermore, a solvent cleaning step just after reflow soldering may also be necessary if some of the conventional components are solvent-sensitive, which would require a final aqueous cleaning process. To avoid two different cleaning processes, when some conventional parts are solvent-sensitive, the solder paste should have water-soluble flux instead of a solvent-soluble one. This still leaves another concern in that there are very few, if any, good solder pastes with water-soluble fluxes. Pathway B is the shortest and least expensive of the two possible pathways and is typically used when the assemblies are manufactured by in-line or highly automated processes. In this case, a cleaning process is avoided after reflow soldering and is only performed just after wave soldering. In the event that pathway B is taken and IR reflow instead of vapor phase reflow is used, a careful cleaning analysis should be performed to determine if acceptable assembly cleanliness levels can be reached. The concern here is that IR reflow tends to decompose more rapidly and bake on the flux residues to the PCB, whereas the vapor reflow process usually rinses off a major portion of the flux residues and thereby imposes significantly less flux residue removal during the cleaning process.

There is still another possible manufacturing sequence for a type I design which incorporates SMDs with gull-wing leads only, along with conventional topside components. This sequence can be shown as follows:

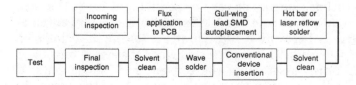

This type of manufacturing process, as shown, avoids the use of solder pastes and reflow solders SMDs via either hot bar or laser reflow. These two types of reflow process are explained in Chap. 10. When these reflow processes are used, the same machine normally places the SMD and reflows it simultaneously. Due to this technique, SMD placement and reflow takes about 6 s per component. High pin count components or those components with lead centerline spacings below 0.040 in. are commonly assembled and soldered in this way; as the pin count increases, the time taken to place, align, and reflow increases.

Design type II, which has surface mount ICs on the top side with conventional components and surface mount chip or discrete components on the bottom side, has a manufacturing process sequence as follows:

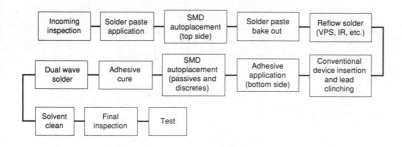

As shown in the above sequence, the conventional components are inserted after the top SMDs have been reflowed but before the adhesive bonding and placement of the SMDs on the bottom side of the PCB. This is a common practice and is done in order to prevent potential damage to the SMDs on the bottom side due to lead clinching. Furthermore, insertion of conventional components usually requires some force, which can lead to a large enough mechanical shock to cause the bonded SMDs on the bottom to fall off the PCB. If the conventional components are inserted after bonding the SMDs onto the bottom side, extreme care should be exercised during lead clinching. Furthermore, the bonding adhesive used to attach the bottom-side SMDs should have very high adhesion strength in order to survive the mechanical shock imposed during conventional component insertion and for the additional handling time of the assembly prior to dual-wave soldering.

Design type III, which can be considered the most complex type of SMA, consists of surface mount ICs on both sides of the PCB along with passive and discrete components. The manufacturing process sequence for this type of SMA can have two possible pathways as follows:

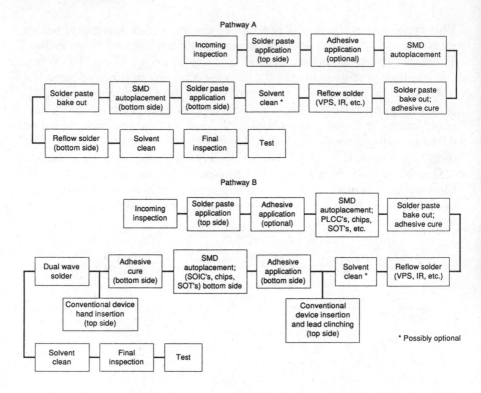

Pathway A

Pathway B

* Possibly optional

Pathway A sequence is necessary when PLCCs are mounted on both sides of the PCB. Whether composed of J or gull-wing leads, these components are currently incapable of being dual-wave-soldered without the occurrence of solder shorts. Therefore, both sides of the assembly will require solder paste printing and reflow solder such as vapor phase or IR. Furthermore, since the top side will be exposed to the reflow process twice, it may be necessary to bond these SMDs to the PCB. This can be avoided by restricting its exposure to the reflow soldering temperatures, which can easily be done with IR reflow. However, this is much more difficult when using vapor phase reflow. In this event, processing the assembly so that the tops of the PLCCs or ICs rest on a flat surface will prevent these components from falling off. Other SMDs such as SOICs and chips in most cases will hold themselves to the PCB due to the combined surface tension of the solder if it becomes molten. If no adhesive is used on the top side, there is always a chance that slight vibrations from the conveyor belt or processing platform may cause components to fall off, misalign, or produce disturbed solder joints due to component movement just prior to solder solidification.

Pathway B sequence, for type III assemblies, can be used when the only ICs placed on the bottom side of the PCB are SOICs. Since these components can be effectively dual-wave-soldered, solder paste printing to the bottom side can be eliminated and dual-wave soldering can be used for the components on the bottom side, and conventional components mounted on the top side.

7.2 Documentation

Documentation will play an important part in anyone's success when it comes to manufacturing SMAs. Basically there are two types of documentation—drawings and specifications. Although these are words familiar to all of us, the proper use, incorporation, and establishment of these documents throughout the industry still have not reached what can be considered satisfactory levels.

Often purchasing is instructed to order components, PCBs, assemblies, and other items requiring subcontracting but is given insufficient or inadequate information or requirements. Furthermore, when subcontracted items are received, the insufficient documentation given to the subcontractor may ultimately lead to unacceptable items. This can consist of a multitude of conditions such as lead tolerances on components, PCB warp and twist, and assembly workmanship. To avoid or to minimize these conditions, which can lead to catastrophic events, it is necessary first for us to understand the wide spectrum of documents required for successful manufacturing of SMAs.

Figure 7.1 illustrates the spectrum of documents necessary to control the manufacturing of SMAs. As shown, there are four major

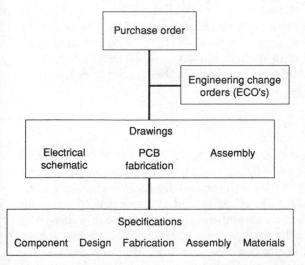

Figure 7.1 Documents necessary for the control of manufacturing SMAs.

groups of documents including the purchase order, engineering change orders (ECOs), three types of engineering drawings, and five types of manufacturing specifications. In some cases, the assembly or other portions of manufacturing may be accomplished in-house instead of subcontracting. When this occurs, a document such as a work order may take the place of a purchase order which allows the execution of the work required.

Today there are several fine organizations such as the EIA and IPC which provide many types of component soldering and assembly specifications. These commercial specifications and military specifications are normally written by committees made up of individuals from a variety of different companies. Often the final specification will end up with a significant amount of information which may not be necessary or applicable and, in some cases, may not be consistent with the requirements actually needed. We must all realize that specific requirements imposed on building, let us say a component or an assembly, may lead to a cost increase and in some cases a cost decrease. However, as the requirements increase and have less manufacturing tolerances, we all can expect the costs to increase. Therefore, it is very important to carefully review any type of commercial document which may be specified and when necessary amend such document when exceptions are taken.

Currently, many companies find it necessary to write their own in-house documents based on their own needs; some of these turn out to be part of commercially available documents and other requirements. In the event that one needs to establish some in-house documents for manufacturing SMAs, let us review some of the basic requirements which should exist as a minimum in these documents. The following sections will review these concerns as a starting point for those who need to initiate a specific document. However, keep in mind that the first step should be to obtain and review all the available documentation, both commercial and military.

7.2.1 Purchase order

The purchase order is a document we are all familiar with. It is used for subcontracting deliverable items and specifies the quantity, types, price, and delivery date of the deliverable items. What is not commonly known about a purchase order is that it takes precedence over all other documentation such as engineering and assembly drawings or specifications. The purchase order should always list the main specifications and drawings of the SMD or assembly that the deliverable item assembly must conform to. If there are special conditions that must be performed on the deliverable item but that are not specified in any document given to the subcontractor, this special condition can be specified on the purchase order. If this occurs, sometimes the incorporation of the special condition in the end product could be simply forgotten by the subcontractor since the purchase order is usually separated from engineering and manufacturing documents. When this occurs, this special condition never gets known to manufacturing, who ultimately can ship the end product out of specification. To avoid this problem, it is always best to use engineering change orders (ECOs), which are discussed in the next section.

7.2.2 Engineering change orders (ECOs)

An ECO is a document which is used to make changes to any existing documents that are specified. Prior to release, the ECO should always be signed off by design and manufacturing engineering, as a minimum, to prevent possible adverse effects to product performance or manufacturing. In most cases, ECOs are used to implement changes such as jumper wires, component values, and even such things as assembly serialization. Whatever is changed or implemented during the manufacturing cycle, we must all understand that verbal orders should never take the place of ECOs. A good practice is to avoid verbal orders.

7.2.3 Engineering drawings

There are three main types of engineering drawings: the electrical schematic, the PCB fabrication, and the assembly. The electrical schematic drawing provides the electrical circuitry configuration of the PCB assembly. It should also provide a list of component types and their manufacturer's part number. A common problem with electrical schematic drawings is that they are often received by the PCB designer or subcontractor with hand-written modifications. In most cases, designers will accept these types of changes, provided they are initialed and dated. However, if such changes are component types, it would always be a good precaution to verify with purchasing whether the new component change is acceptable based on pricing and availability. Still the best procedure is to submit changes via an ECO which has been reviewed and approved by all concerned.

The PCB fabrication drawing should specify all the necessary materials, construction, and tolerances the fabricator must conform to. In all cases, a fabrication document such as MIL-P-55110 or IPC-D-320 should be specified as a minimum. Furthermore, when materials are specified, it is a good practice to call out a specification which controls that material. For example, a common practice is to just specify "use liquid solder mask" in the fabrication drawing notes. The best approach is to specify a liquid or photoimagable solder mask, the brand (i.e., PC401, SR1000, Vacrel, etc.), and a requirement specification such as IPC-SM-720.

The assembly drawing illustrates the layout of components and their locations. It also provides other assembly requirements such as adhesive application and component mounting conditions. In all cases, the assembly drawing should specify an assembly requirements specification.

7.2.4 Specifications

The necessary specifications, at a minimum, required to manufacture SMAs consist of design, component, PCB fabrication, manufacturing materials, and assembly specifications. There are two types of specifications, those that specify "what" and those that specify "how," which are more commonly called procedures or guidelines. In most cases, when subcontracting is performed, the subject specifications should provide what is required. Once the subcontractor knows the specified requirements, then internal specifications are used by the subcontractor to describe how the product is to be manufactured. In some cases, the specifications which provide end product requirements will typically limit the manufacturer to a specific condition, such as a solder alloy or substrate material, but not require a specific brand name. When it comes to manufacturing SMAs, these five specifications should be carefully established and reviewed by all those involved. When conditions are not specified or are specified improperly, do not expect that the manufacturer will provide you with exactly what you want, even though the manufacturer may have provided initial lots of product in conformance with your expectations.

7.3   Incoming Inspection Process

In manufacturing SMAs, incoming inspection will become a matter of survival in the electronic industry. Currently, it may be surprising to hear that most companies do little, if any, incoming inspection of their components, materials, or PCBs prior to releasing them for manufacturing assembly. Even in the infancy stages of SMT being used in the commercial industries, many stories can be told about massive chip component defects, component solderability problems, manufacturing material inconsistencies (solder pastes, fluxes, etc.), and a variety of PCB problems such as solder mask delamination and excessive board warp and twist.

The major goal of incoming inspection is to find defective or unacceptable materials prior to their release to manufacturing. Over time, the quality levels of each material tested can be analyzed and fed back to the manufacturer of the material so as to aid in increasing the quality levels to a point where incoming inspection becomes unnecessary. Incoming inspection can also be a starting point for the qualification of a new vendor's product.

Incoming inspection of components, materials, and PCBs for high-volume manufacturing of SMAs should actually become part of the process. In this circumstance, finding such things as component solderability defects after potentially thousands of these defective components have been placed and soldered could cause crippling results for any company. Therefore, we must single out the most likely problems during the process of incoming inspection, but definitely not waste time and money in areas in which problems occur infrequently or are less likely to occur.

7.3.1 SMDs

There are three major types of SMDs that are the basis of this discussion of incoming inspection: PLCCs (J and gull-wing leaded), chip resistors, and chip capacitors. The main issues of concern are lead coplanarity, solderability, and the external and internal workmanship conditions of both types of chips. Let us now review these concerns in the following sections.

7.3.1.1 Lead coplanarity.

While completely coplanar leads on PLCCs are desirable, no SMD manufacturer will be capable of supplying components with 100 percent of the leads coplanar. The major cause of the problem is improper packaging techniques and handling. For this reason, the JEDEC Solid State Products Engineering Council has issued a standard tolerance value that has satisfied both manufacturers and users. The standard specifies that the bottom of the leads shall fall within a 0.004-in.-wide tolerance zone.[1] This tolerance zone is established by two parallel planes that are determined by the three lowest points of the leads, that is, the three points upon which the PLCC can rest in a stable position. The coplanarity measurement can be made in several ways; however, the easiest method is simply placing the PLCC on an optical flat and using a toolmaker's microscope to visually measure the distance between non-coplanar leads and the seating plane. Furthermore, we must realize that incoming inspection should not be merely a process to collect information or to run irrelevant tests but to find and set conditions which ultimately will affect manufacturability or reliability. It is quite clear that lead coplanarity can have adverse affects on manufacturing yields and solder joint reliability. When leads are excessively non-coplanar, they cause solder opens, which can be difficult to observe. Also, experimental tests have shown strong evidence that the greater the solder bridging distance between the lead and land, the weaker the solder joint will be.[2]

7.3.1.2 Solderability. The solderability of component terminations (leads, end caps, etc.) is the major condition leading to either high or low soldering yields. Solderability is a measure of the ability of a metal surface to be wetted by molten solder and to stay wetted. It is important to recognize that the solderability of a component is not an absolute property but depends on such soldering conditions as time, temperature, flux activity, and type of lead metallization.

There are several causes of poor solderability; however, most assume that the cause is simply oxidation. In fact, it is well established that if a component is made solderable with a thick layer (0.005 in. or more) of a solderable coating, such as electroplated tin or solder, it will remain solderable for a significant period of time. Thus, the conception of a layer of oxide on the surface of the coating somehow totally preventing solderability is typically not the correct answer if the problem occurs. This is because the thin oxide layers which develop over lead coatings during storage break up readily due to the chemical action of the flux and the physical effect of the melting of the underlying coating. With this understood, we can better comprehend the major cause of most of our component solderability problems. This is an insufficient coating thickness over the base metal, which is typically nickel plated over copper or alloys of iron. Although the solderability of the component with a thick coating may test acceptably soon after plating, it can deteriorate, primarily due to solid-state diffusion and chemical reaction with the underlying base metal, which forms intermetallic compounds such as $NiSn_3$ and $NiSn_4$. If the nickel barrier layer over the base metal is thin or missing in areas, then other intermetallics can form such as Cu_3Sn or Cu_6Sn_5 for copper leads and $FeSn_2$ for iron alloy leads. The rate of these possible reactions increases exponentially with time and temperature. Furthermore, this reaction uses up the tin coating until in some cases none remains and the intermetallic compound oxidizes at any spot where it becomes exposed. Once the intermetallic compound oxidizes, it is highly unlikely that even very activated fluxes will reduce it such that it becomes solderable. Figure 7.2 illustrates three separate conditions which demonstrate the effects of tin coating thicknesses over nickel-plated leads.

Chip components such as resistors and capacitors have their own set of solderability problems other than the one mentioned above. Chip resister manufacturers normally plate over nickel with a very high

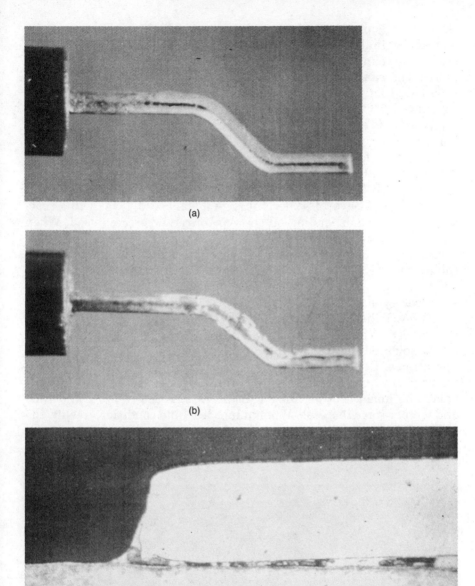

(a)

(b)

(c)

Figure 7.2 Solderability conditions of various types on nickel-plated, tin-coated leads. (a) Good solderability or wetting; (b) poor solderability or dewetting; (c) cross section of a gull-winged lead soldered to a land (pad) showing solder nonwetting to lead creating voids and a reduction in reliability.

tin-lead ratio plating such as 90/10 or 95/5. The high tin content, designed originally for wave soldering operations, tends to be less solderable than tin-lead ratios of 72/28 when reflow-soldered via vapor phase. The exact cause for this is still unclear; however, it has been reported that the lower tin-lead ratio will aid the ease of solderability.[3] Chip capacitors are commonly supplied with end caps composed of either palladium silver or palladium silver–nickel barrier–electroplated tin. The end caps composed of palladium silver are notorious for producing a phenomenon called *leaching* when soldered with tin-lead–bearing solders. When this occurs, the tin in the solder readily reacts with the silver, producing brittle silver-tin intermetallics. Several solderability problems result when this reaction takes place depending on the type of reflow process used. When dual-wave soldering is used, as a result of the scrubbing action of the turbulent wave, large amounts of these intermetallics can easily dissolve completely off the chip itself. Figure 7.3 illustrates this condition when a chip capacitor with palladium silver end caps is wave-soldered. Typically, the palladium silver on the corners and edges of the chip will dissolve away first since these are the thinnest regions of the end cap. When the same type of chip capacitor is vapor phase or IR reflow-soldered, leaching will occur; however, it will take a different appearance. Depending on the length of dwell time at the liquidized solder phase and the reflow temperature, the amount of leaching will vary. The amount of leaching will increase as the reflow dwell time and temperature increase. When minor leaching conditions result, the solder fillet can still be properly shaped but will have grainy or "bubbly" appearances near the thinnest regions of the joint. When the

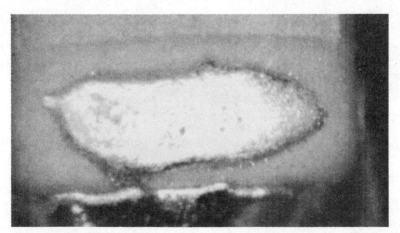

Figure 7.3 Dissolution of the palladium silver end cap on a chip capacitor after being dual-wave-soldered.

Figure 7.4 Cross section of a chip capacitor with severe leaching of the end cap. Also, notice the complete dissolution of the palladium silver end cap from the ceramic body leaving a black voided region.

most advanced stages of leaching occur, the solder fillet will have high concentrations of the intermetallics, causing it to actually clump to the end caps as shown in the cross section in Fig. 7.4. Allowing minor leaching conditions may result in solder joint cracking over time. Reworking any type of leaching condition will only advance the chemical reaction, producing more intermetallics, and in most cases only mask the connection with a visually good solder joint but a very unreliable internal joint.

Understanding what our major component solderability problems are leads us to the next step in determining how to measure component solderability. Solderability can be measured qualitatively and quantitatively.

Qualitative solderability measurement is performed by simply dipping the component terminations into a bath of molten solder and then removing them for visual evaluation. There are several specifications which govern this solderability test, as shown in Fig. 7.5; however, the basic test consists of dipping the sample in either an activated or nonactivated flux and removing any drops of excessive flux. After this, the sample is dipped not less than 2 mm below the molten 60/40 or 63/37 Sn-Pb solder bath to an immersion time of from 20 to 25 mm/s. This part of the so-called dip test method is best controlled by a dip tester machine, as shown in Fig. 7.6. The sample immersion dwell

Figure 7.5 Qualitative Solderability Test Methods

Specimen	Pretreatment	Solder	Flux	Temperature, °C	Procedure	Evaluation
IEC 68-2-20 Ta Method I						
Wires, tags, terminations of irregular form	As globule test, see Fig. 7.7	Min. 300 cm², Sn60-Pb40	Immerse specimen in flux and drain for 60 s; compositions as for globule test (Fig. 7.7)	235 ± 5	Dip specimen at 25 ± 2.5 mm/s; dwell at specified depth for 2 or 5 ± 0.5 s depending on heat capacity	Small, scattered faults permitted in solder coating viewed up to 10×
EIA RS-178-B						
Terminals up to 0.045 in. (0.114 cm)	None permitted	Min. 112 cm², Sn60-Pb40	25% w/w colophony in isopropanol; dip for 5–10 s in flux and dip in bath 15–20 s later	235 ± 5 or 271 ± 5	Dip specimen at 25.4 ± 6.4 mm/s; dwell at specified depth for 5 ± 0.5	Scattered faults not exceeding 5% of area allowed, viewed at 10×
EIA RS-319-A Vertical dip						
PCB without PTH; max. dimension, 2 in. (5 cm)	None permitted	Min. 375 cm², 60 or 63 Sn	35% w/w colophony in isopropanol; may have 0.9% Cl on solids	260 ± 5	Dip at 2 ± 6 mm/s; dwell at specified depth for 3–4 s	Scattered faults not exceeding 5% area viewed 2× to 10×
IPC-S-801						
PCB	None permitted	Min. 280 cm³, 60 or 63 Sn	25% w/w colophony in isopropanol	232 ± 5	Dip at 25 ± 6 mm/s; dwell at specified depth for 4 ± 0.5 s	Scattered faults not exceeding 5% area viewed up to 10×

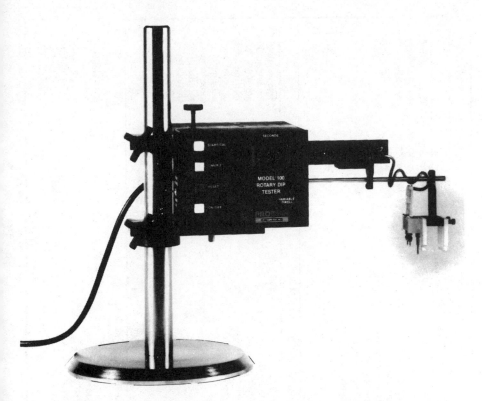

Figure 7.6 An automated dip tester machine used to lower and dip the sample in a molten bath of solder by preset conditions. (*Courtesy International Range Corp.*)

time can vary depending on which specification is used; however, it is strongly recommended that the dwell time be approximately twice the time the components will see in actual production. Although subjective, this type of test is the least expensive way of finding severe component solderability problems, especially leaching conditions associated with chip resistors and capacitors.

Quantitative solderability measurement can be performed by either the globule or wetting balance test methods. Both of these tests are outlined in Fig. 7.7. The globule test method was originally designed to test cylindrical wire by placing the fluxed wire horizontally above a globule of solder, quickly lowering the wire into the globule, and taking the time for the two halves of the globule to close together over the wire. The size of the globule is chosen so that closure cannot occur unless the wire wets to the solder globule. Although it can be performed by an automated machine, this test has not been designed for the small leads, end caps, or metallizations encountered with SMDs and,

Figure 7.7 Quantitative Solderability Test Methods

Specimen	Pretreatment	Solder	Flux	Temperature, °C	Procedure	Evaluation
			Globule Test, Ice 68-2-20 Ta, Method 3			
Wire, circular section; max. distance 1.2 mm	Wires may be detached from component; no cleaning except immersion in neutral organic solvent; aging permitted: steam, 40°C damp heat, 155°C dry heat	Sn-Pb40; globule weight (for given wire diameters): 1.2–0.75 mm, 200 mg; 0.74–0.55 mm, 125 mg; 0.54–0.25 mm, 75 mg; +0.25 mm, 50 mg	25% w/w colophony in isopropanol or ethanol; addition of activator up to 0.5% Cl on solids if appropriate	235 ± 2	Wire bisects globule, which later unites above it	Time required for globule to reunite
			Wetting Balance Test, IEC 68-2-54			
No restrictions	No cleaning except immersion in neutral organic solvent	Sn60-Pb40; specimen no closer than 15 mm to bath walls or base	As 9.1; drain excess flux on filter paper for 1–5 s	235 ± 3	Suspend specimen with lower edge 20 ± 5 mm above bath surface for 15–45 s before immersion at rate 20 ± 5 mm/s; depth to be specified between 2 and 5 mm	Time required for contact angle of 90° for wetting force to equal buoyancy or for force to reach a specified fraction of the maximum
			MIL-STD-883B, Method 2022			
Rectangular terminations, max. 0.050 × 0.025 in. (1.27 × 0.64 mm)	No precleaning permitted; min. 60 min steam-aging mandatory	Approx. 93 cm², Sn60-Pb40	Type R or type RMA of MIL-F-14256; immerse in flux for 5–10 s	260 ± 10	Dip at 25.4 ± 6.4 mm/s for 5 ± 0.5 s at depth of 4 mm	Wetting force equal to buoyancy within 0.59 s and to 300 dyn/cm within 1 s

Figure 7.8 A meniscograph solderability machine used to perform the wetting balance test method. (*Courtesy Multicore Solders.*)

therefore, will not be a common method used for quantitatively determining the solderability of SMDs.

The wetting balance method, which utilizes a machine called a meniscograph, is becoming an increasingly popular method for quantitatively determining the solderability of both leaded and leadless SMDs. The meniscograph, as illustrated in Fig. 7.8, holds the sample with the lead(s) or end cap downward above a solder pot. The mechanism holding the sample is connected to a balance which converts weight to an electrical analog signal. During the procedure the molten solder in the pot is brought up around the lead(s) or end cap, which causes a change in weight and, therefore, a change in electrical signal which can be tracked on a chart recorder. Basically, this method tests both the surface wetting forces of the terminations. As specified in MIL-STD-883, method 2022, and illustrated in Fig. 7.9, certain conditions as plotted by the chart recorder during the meniscograph testing must be met for acceptable solderability conditions. For acceptable conditions, the wetting force must cross the zero balance line in less than 0.6 s, and at 1.0 s the wetting force (F_1) must be at least two-thirds of the maximum wetting force (F_2). A variety of solder wetting conditions can be illustrated in Fig. 7.10 from meniscograph testing. In all cases the initial charted wetting conditions will produce negative nonwetting forces until the termination comes up to solder wetting temperatures. Once the termination reaches wetting temperature and the flux performs, wetting will increase, producing a positive force of various levels depending on the degree of solderability. Nonwetting conditions never reach a positive wetting force.

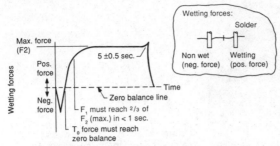

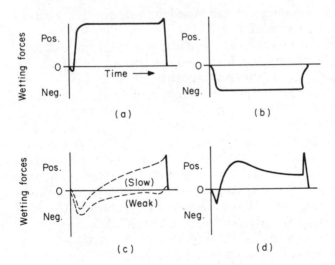

Notes: (1) This test measures both speed and forces of solder wetting during a tightly controlled solder dipping test
(2) Test acceptance is as follows:
-- Wetting force must cross the zero balance line in <0.6 second
-- At 1,0 second the wetting force (F_1) must be at least 2/3 of the maximum wetting force (F_2)

Figure 7.9 Wetting balance solderability test parameters as specified in MIL-STD-883, method 2022.

Notes: (1) Meniscograph wetting 'speed' will be very dependent on thermal conductivity and actual solderability of lead
(2) The wetting 'force' will be dependent on lead surface perimeter and degree of lead wetting

Figure 7.10 Various solderability wetting conditions as a result of wetting balance meniscography testing. (*a*) Rapid wetting; (*b*) nonwetting; (*c*) slow or weak wetting; (*d*) dewetting.

7.3.1.3 SMD workmanship.

The external and internal workmanship conditions of SMDs will be an important part of incoming inspection in order to find defective or marginally reliable components. Inspecting the external workmanship conditions can be simply accomplished under magnification of 10 × and greater, along with measuring techniques done manually or optically. Inspecting internal workmanship conditions will require microsectioning techniques which can be performed in accordance with IPC-TM-650, method 2.1.1, or by other published methods.[4] Microsectioning techniques should be performed by well-trained and experienced technicians since the process of component microsectioning can itself produce internal defects if not done correctly.

There are many types of surface mount ICs which are discussed in Chap. 3. For the purpose of incoming inspection, we can separate surface mount ICs into two different types, leadless and leaded. The leadless types, which are predominantly ceramic leadless chip carriers (LCCs) and are widely used in military applications, have the common defects of layer-to-layer misregistration and layer edge undercutting. The LCC consists of three separate ceramic layers called the base external contact layer, the internal bond pad layer, and the seal ring layer. The bottom two layers are connected to the wraparound constellations or metallizations which make contact to the solder joint. When these two layers are misregistered or have undercut edges, as shown in Figs. 7.11 and 7.12, they can produce highly concentrated areas of stress within the formed solder joint. When the joint is stressed envi-

Figure 7.11   Cross section of a ceramic LCC with layer-to-layer misregistration.

Figure 7.12 Cross section of a ceramic LCC with severe layer edge undercutting.

ronmentally or via power cycling, these added areas of stress can cause a significant reduction in the cycles leading to solder joint failure. Currently, these are no standards which specify the acceptable or reject criteria for these two workmanship conditions; however, Fig. 7.13 can be used as a guideline until the industry analyzes these conditions and determines the extent of their effect on solder joint reliability.

Leaded SMDs such as SOICs and PLCCs have two common external workmanship conditions which should be reviewed at incoming inspection. These consist of component dimensional tolerances and extraneous body material. The component dimensional tolerances are important since the same part number from two different manufacturers can have significantly different body dimensions which may not fit the actual land pattern design on the PCB. Furthermore, when these types of components are fed onto the placement machine via vibratory or gravity feeders, small variations from lead-to-lead distances can cause components to jam. Not being able to control component lead-to-lead distances to 0.010 in. may force one to package these components in tape and reel. Another common external workmanship condition found on leaded SMDs consists of extraneous body material, which can affect the alignment of the component with the PCB if mechanical alignment methods are used. Extraneous body material is caused during the mold injection process of the epoxy resin and usu-

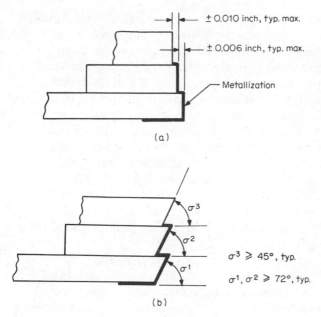

± 0.010 inch, typ. max.

± 0.006 inch, typ. max.

Metallization

(a)

σ^3

σ^2

σ^1

$\sigma^3 \geqslant 45°$, typ.

$\sigma^1, \sigma^2 \geqslant 72°$, typ.

(b)

Figure 7.13 Internal workmanship guidelines for ceramic LCCs. (*a*) Layer-to-layer misregistration; (*b*) layer edge undercutting.

ally occurs between the component leads, as shown in Fig. 7.14. Reject conditions associated with extraneous body material should simply be any amount of material which makes contact with the squaring mechanisms during the alignment stage of the component, which causes unacceptable component misalignment. When vision alignment is used, this condition no longer has any importance unless component-to-component spacing is small.

Figure 7.14 Extraneous epoxy encapsulation material protruding from the side of the SMD and in between its leads.

Internal workmanship conditions for leaded SMDs should include the lead plating construction and the lead-to-body interface integrity after the component has been immersed in molten solder representative of the dual-wave soldering process, if used. After microsectioning is performed, the above workmanship conditions can be determined. The typical lead plating construction consists of a base metal, a nickel-plated layer, and an electroplated tin or solder coat layer. Each metal layer should have a uniform thickness consistent with the requirements and show no signs of delamination or excessive voids. Furthermore, after immersion in solder and cross-sectional viewing, the component should show no signs of lead-to-body delamination or solder wicking between the lead and body material. As mentioned, this test and inspection is only necessary when the component is dual-wave-soldered. Figure 7.15 can be used for workmanship guidelines to the internal conditions of leaded SMDs.

Chip components, both resistors and capacitors, have a wide variety of external and internal workmanship conditions which should be verified and examined at incoming inspection. These components, although small, are more complex in their construction and materials than most expect. However, these components produce a greater number of component failures at the assembly level than any other type of SMDs. The following workmanship guidelines can be used for chip resistors.

External visual workmanship conditions
Major defects

1. Foreign debris in resistive film which exceeds 0.005 in. in its longest dimension.

2. Alumina substrate with large shaped alumina particles in excess of 4 μm (see Fig. 7.16).

3. Rough substrate edges which are visually noticeable (see Fig. 7.17).

4. A laser or mechanical trim path length that exceeds 50 percent of the width of the resistive film.

5. A laser or mechanical trim cut depth or kerf less than 10 μm.

6. Excessive microcracking in the resistive film caused by laser or mechanical trim process.

7. Evidence of resistive film debris in laser or mechanical trim path.

8. Fine line fracture in substrate.

9. Overglaze passivation (lead borosilicate glass) which covers more than 25 percent of the end cap width.

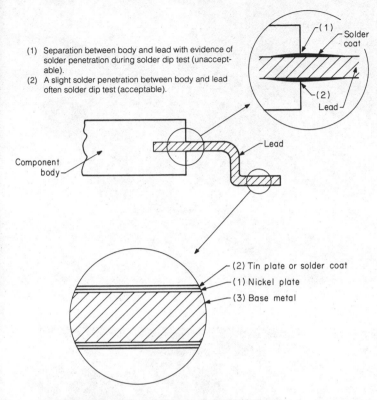

(1) Separation between body and lead with evidence of solder penetration during solder dip test (unacceptable).
(2) A slight solder penetration between body and lead often solder dip test (acceptable).

Note: (1) Nickel plate uniform in thickness with good evidence of wetting and specified thickness.
(2) Tin or solder plate uniform in thickness and complete wetting to nickel plate after solder dip. Specified thickness prior to solder dip.
(3) Base metal uniform in thickness with clean surface conditions and specified thickness.

Figure 7.15 Workmanship guidelines for the internal construction of leaded SMDs.

Minor defects

1. Foreign debris in resistive film which is less than 0.005 in. in its longest dimension.

2. Alumina substrate with small rounded alumina particles (1 to 2 μm) (see Fig. 7.16).

3. Visually undetected substrate edge roughness.

4. A laser or mechanical path length less than 50 percent of the width of resistive film.

5. A laser or mechanical trim cut depth or kerf greater than 10 μm.

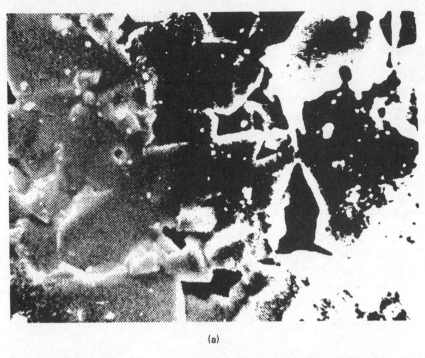

(a)

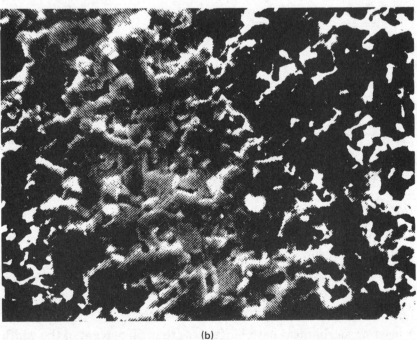

(b)

Figure 7.16 Alumina substrate on chip resistor. (*a*) With large shaped alumina particles in excess of 4 μm; (*b*) with small rounded alumina particles, 1 to 2 μm.

Figure 7.17 Rough substrate edges on a chip resistor caused by laser cutting.

6. Overglaze passivation which covers less than 25 percent of the end cap width leaving no exposed resistive film.

Internal workmanship conditions
Major defects

1. Irregular resistive film thicknesses which could affect resistor performance.
2. End cap contact with the resistive film of less than 95 percent of the width of the end cap.
3. Thicknesses of the inner electrode, nickel barrier, and thin lead coating less than the specified amount.

Minor defects

1. Some irregularity in the resistive film but not significant enough as to affect performance.
2. End cap contact with the resistive film greater that 95 percent of the width of the end cap.

The following workmanship guidelines can be used for chip capacitors.

External workmanship conditions (see Fig. 7.18)
Major defects

1. Cracks, blisters, or delaminations in capacitor body.
2. Chips or voids in the capacitor body which exceed 0.003 in. or which expose internal electrodes.

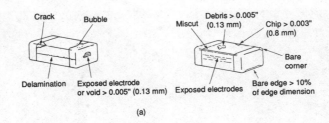

(a)

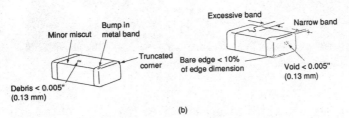

(b)

Figure 7.18 Classification of external workmanship conditions for chip capacitors. (a) Major defects; (b) minor defects.

3. Foreign debris bonded to the chip surface which exceeds 0.005 in.

4. Miscuts of the capacitor edge which penetrate the surface by more than 0.005 in. or expose internal electrodes.

5. End cap defects consisting of: (a) Voids in the end cap with exposed electrodes; (b) voids in the end cap exceeding 0.005 in.; (c) exposed metallized edges which exceed 10 percent of the edge dimension; and (d) bare corners or metallized ends.

Minor defects

1. Minor irregular cuts of sides or corners with no exposed electrodes.

2. Foreign debris on capacitor surface not exceeding 0.005 in.

3. End cap defects consisting of: (a) Voids in metallization less than 0.005 in.; (b) exposed metallized edges which do not exceed 10 percent of edge dimension; (c) excessive metallization which reduces the gap between end caps by more than 30 percent; (d) insufficient metal end cap dimensions which do not affect solder joint attachment; and (e) excessive metallization which prevents chip capacitor from lying flat.

Internal workmanship conditions (see Fig. 7.19)
Major defects

1. Delamination or separation between the electrode and ceramic, or within the ceramic dielectric, associated with warpage of the elec-

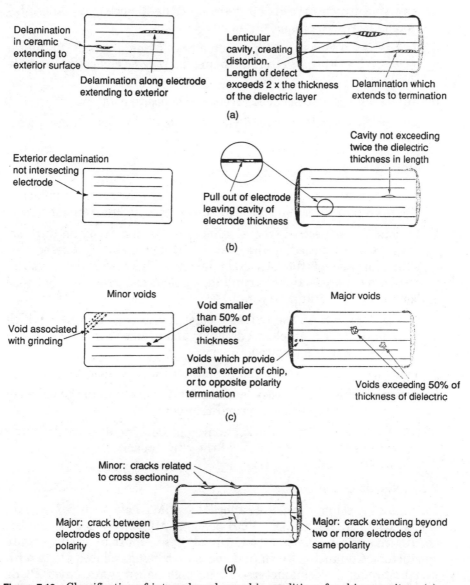

Figure 7.19 Classification of internal workmanship conditions for chip capacitors. (*a*) Major delaminations; (*b*) minor delaminations; (*c*) void criteria; (*d*) crack criteria.

trode layers, and surrounding layers exceeding two times the dielectric thickness dimension.

2. Delamination or separation of ceramic and electrode or of layers within the dielectric which extends to the exterior side of the chip capacitor, or to an electrode or end cap of opposite polarity.

3. Any void which exceeds 50 percent of the dielectric thickness between electrodes of opposite polarity.

4. Any series of voids which provide a path along a surface to the exterior of the chip capacitor, or to an electrode or end cap of opposite polarity.

5. Any crack which bisects electrodes of opposite polarity.

6. Any crack extending beyond two or more electrodes of same polarity.

Minor defects

1. A delamination along an electrode or within the ceramic dielectric of length less than two times the dielectric thickness dimension.

2. Pullout of the electrode metal which results in a cavity on the electrode axis not exceeding the electrode thickness and not associated with warpage of the electrode layers. (This defect is normally caused by cross-sectional grinding, as electrode layers are not well bonded to the ceramic layers.)

3. A delamination on the exterior edge of the chip capacitor which does not intersect an electrode layer and is less than 50 percent of the distance to the electrode layer.

4. Voids which are less than 50 percent of the dielectric thickness.

5. Series of voids clearly associated with grinding lines in the polished section attributed to the sectioning process.

6. Cracks associated with the sectioning process, notably cracks along the cover sheet ceramic layer of the chip capacitor.

7.3.2 Printed circuit boards

The higher assembly densities afforded by SMT have resulted in more complex and higher density PCB designs. Multilayer boards are becoming more frequent since additional layers must be used to route conductors. Conductor width and spaces are being reduced along with plated through-hole diameters. These high-density multilayer PCBs and even the two-sided high-density PCBs are becoming more of a challenge for most of the PCB fabricators, which has resulted in increasing the possibilities of receiving PCBs that are unacceptable or marginally acceptable and possibly unreliable.

An interesting factor about PCBs, especially those that see SMT manufacturing assembly processes, is that they must actually be fabricated with more reliability or durability since SMT assembly processes can impose significantly greater amounts of thermal stresses than conventional through-hole technology manufacturing. Therefore, incoming inspection of PCBs should be performed in such a way as to uncover potential defects or reliability problems. It should be mentioned that even if a PCB "looks good," it still may not be a reliable one or one that may even meet the requirements after or during its exposure to the manufacturing assembly processes. The key issues, as discussed in the following sections, concern the most important areas of the PCB that should be examined and tested at incoming inspection. It should also be pointed out that incoming inspection of PCBs, which consists of visual inspection only, will not uncover latent defects such as excessive warp and twist, PTH integrity, and solder mask integrity, which may only surface during the manufacturing processes or incoming inspection tests representative of the processes.

7.3.2.1 PCB dimensional measurements. The PCB dimensional measurements should be verified prior to lot acceptance since factory board fixturing, such as for screen printing and component placement, may have fixed dimensions. The basic concern is simply to measure the tooling hole diameters and their distances from each other. Sometimes the board edges are used for alignment purposes instead of tooling holes; therefore, it is necessary to measure the board's perimeter dimensions for conformance. One should keep in mind that no matter how high the quality of a PCB is, it makes no difference if the PCB cannot fit the manufacturing tooling necessary to build the assembly.

7.3.2.2 External workmanship. There are numerous types of external workmanship conditions which should be inspected at incoming. As a minimum, inspect for the following conditions:

1. Solder mask flow onto the lands, or, when photoimagable solder masks are used, their alignment with respect to the lands.
2. Solder mask delamination, foreign matter entrapment, or, when dry films are used, wrinkling, poor encapsulation of conductors, and delamination.
3. Annular ring widths.
4. Conductor etch resolution width tolerances.
5. Laminate delaminations, halowing, measling, and weave-fabric exposure.

7.3.2.3 Destructive external analysis. Destructive external PCB analysis at incoming inspection is performed in order to catch any latent defects which could occur only during the manufacturing assembly processes such as soldering and cleaning. The three main areas of concern are PCB warp and twist, solderability, and solder mask integrity.

PCB warp and twist, of course, can be measured without destruction to the PCB or lamination process (for multilayer PCBs). PCB design can cause warp and twist by large variations in copper and glass regions on the PCB, concentrations of areas with no PTHs or with PTHs, and portions of the PCB laminate which are cut out either along the board edges or inside its edges. The fabricator will have little control, if any, when PCB warp and twist is caused by board design. Incoming inspection of board warp and twist can consist of the measuring techniques specified in IPC-TM-650, method 2.4.22; however, prior to any type of measuring technique, the PCB should be exposed to thermal stresses representative of what it will see in manufacturing assembly. The typical thermal stress test, which also can be used to test solderability and solder mask integrity, is the rotary dip test or solder float test where the PCB is immersed in a bath of solder for a specific time. Again, the test should, as a minimum, be representative of the manufacturing process; in most cases the available commercial or military specifications require revision to accommodate this concern.

The solderability of the PCB is another important condition which should be verified at incoming inspection. Of greatest concern are the land areas where the SMDs are soldered and the PTHs. IPC-S-804 specifies the solderability test methods for PCBs, which consist of the edge dip test, the rotary dip test, the wave solder dip test, and timed solder use test. The edge dip test is designed for surface conductors only. The rotary dip and wave solder test are designed for PTHs and surface conductors. The time solder use test (sometimes called the globule test) is designed only for PTHs. A convenient approach to solderability testing of the PCB, regardless of the test used, is to design in a PCB coupon, with various land geometries, several sized PTHs, and at least two large exposed conductor planes.

When used on surface mount PCBs, solder mask is more of a quality concern than ever before. Liquid solder masks, which are widely used for conventional PCBs, are becoming less appropriate for high-density SMAs due to insufficient alignment accuracy and flow characteristics. Dry film solder masks are therefore becoming more popular for use on high-density PCBs along with liquid photoimageable solder masks since they have high resolution characteristics and no flow properties.

Quality concerns, however, do exist with most dry film solder masks. These solder masks are laminated onto the PCB under pressure and heat, which requires efficient processing and extremely clean PCB surfaces. Typically, these solder masks are laminated on bare copper PCBs since their adhesion over tin-lead surfaces, which becomes molten during reflow, is very poor. Their exposure to the more severe reflow processes used in SMT tends sometimes to induce enough thermal stress to cause dry film masks to delaminate from the PCB surface and to crack. Dry films are also known to be very brittle; therefore, they tend to microcrack under thermal and mechanical stresses such as when warped and twisted boards are flattened out. Another latent quality problem with solder masks, common only to fabricators who are inexperienced with the curing and laminations processes involved with dry film solder mask applications, is their physical and chemical breakdown during exposure to cleaning solvents. When cured improperly during lamination or improperly exposed to ultraviolet radiation (for developing reasons), the dry film material may be poorly cross-linked and, therefore, chemically reactive to cleaning solvents. Solder mask flaking or peeling off the PCB is a common result of improperly cured or insufficiently cross-linked film. In order to uncover these potential and serious latent defects in dry film solder masks, incoming inspection tests, both thermal and solvent exposure, should be performed. Thermal exposure should consist of a solder float test for 10 to 15 s with the solder bath temperature between 500 and 550°F. This type of thermal stress test will be more severe than what would be expected for most PCBs during the manufacturing process. During this thermal stress test, delamination or separation of the solder mask from the board surface can occur but is sometimes difficult to observe. For ease of observation, the tested sample should be immersed in water and tapped against its container to promote the capillary action of water between the solder mask and the board surface. If a separation exists between the solder mask and board surface, then entrapped water will be noticeable. Furthermore, after the test sample has been through the thermal stress test, the same sample should be exposed for a minimum of 15 min to the same cleaning solvent used in manufacturing. If a mild cleaning solvent, such as the chlorofluorinated types, is used in manufacturing, it may be wise to use a stronger solvent such as the 1,1,1-trichloroethane azeotropic solvents or blends to uncover potentially marginal solder mask conditions.

7.3.2.4 Internal workmanship.
Examination of PCB internal workmanship conditions requires microsectioning techniques, which can be performed per IPC-TM-650, method 2.2.11. Coupons, which are preferred over destroying an actual PCB, should be cross-sectioned through the PTHs designed in the coupon. Microsectioning should take place after the coupon has been through a solder float thermal stress test. Under magnification some of the most critical areas to inspect for conformance with the requirements consist of the copper and tin-lead plating thicknesses, layer-to-layer misregistration of internal

conductors, resin etchback, resin smear removal, laminate voids, and copper cracks or fractures as they are applicable to two-sided or multilayer boards.

Along with specifications such as IPC-D-320, MIL-STD-55110, and IPC-A-600A, a great deal of information is available on the acceptability of PCBs.[4,5] Since PCB internal workmanship conditions and inspection procedures have been well documented for many years, further discussion of these concerns will not be necessary. However, it is strongly suggested that multilayer PCBs be internally inspected in the plated through-hole area for each lot that is received. Most commercial companies tend not to do this with their PCBs since the majority are used for simple constructions in two-sided PCBs and thus the potential problems that can occur in multilayer PCBs are not expected.

7.3.3 Manufacturing materials

The major types of manufacturing materials used to assemble SMAs are solder pastes, solder alloys, fluxes, and solvents. Each one plays an important part in the quality level of the assembly and the level of assembly defects that can occur as a result of poor material quality and material that is nonconforming, either physically, chemically, or both.

It is important for the user of these materials to realize that, regardless of how well a material may perform initially, its continual performance can vary and sometimes drastically change due to inconsistencies or variations in the material manufacturer's process. Due to these likely variations in manufacturing materials, adequate incoming inspection techniques should be required as they are received. Let us now review these concerns in the following sections.

7.3.3.1 Solder pastes. Solder paste can be the most complex of all the materials used to manufacture SMAs. Typically, there are no two types of solder paste which behave the same, and, more frequently than expected, there are no two lots from the same manufacturer that also behave the same. Solder paste itself is a complex material both physically and chemically. For these basic reasons, it is necessary for the user to first extensively qualify the solder paste which will be used in manufacturing by comparing several types of solder paste formulations from various manufacturers. Once a solder paste is qualified, then incoming inspection of each lot should consist of four different tests, as a minimum. These four incoming inspection tests are the percent metal, solder ball formation, viscosity, and powder oxide content tests. These four types of incoming inspection tests are chosen over others that exist since these are the most likely areas in a solder paste to vary or change considerably.

Percent metal content. The percent metal content or metal powder load in solder paste is a major factor affecting solder paste viscosity and rheology (flow properties). Depending on the application method, solder pastes usually range in percent metal content from 85 to 92 by weight. Solder pastes with higher percents of metal content help to control solder voids, paste slump during bakeout, and excessive flux residues. However, too high a percent metal content can cause difficulties in printing or dispensing techniques, an increase in powder oxidation, and a reduction in solder paste shelf life. A quantitative method for measuring the solder paste percent metal content is as follows:

1. Weigh sample (to the nearest 0.1 g) of solder paste in a crucible (subtract weight of crucible).
2. Heat crucible and solder paste on a hot plate or other reflow method until the solder melts and a complete separation of the flux and metal occurs.
3. Allow metal to solidify, remove flux with suitable solvent, and dry thoroughly.
4. Weigh the metal ball and calculate the percent metal content as follows:

$$\% \text{ metal content} = \frac{\text{weight of metal ball}}{\text{weight of solder paste}} \times 100$$

Solder ball formation. Solder balls are solder particles of various sizes which remain apart from the solder joint during reflow and are unable to coalesce with the molten solder joint while it solidifies. Solder balls are produced as a result of excessive metal oxides on the metal particles and from small particles carried out by the displaced flux during reflow. In addition, solder balls can also form due to solder paste spattering during bakeout and reflow caused by rapid escape of entrapped volatiles. The obvious problem with solder balls is that they can cause electrical shorts if they bridge between adjacent conductors or become dislodged and cause shorting during the use of the product. Figure 7.20 illustrates acceptable and unacceptable levels of solder balls as a result of testing solder paste for this condition using the following test method:

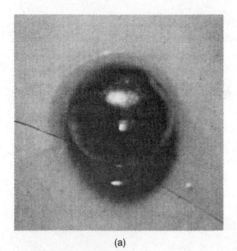

(a)

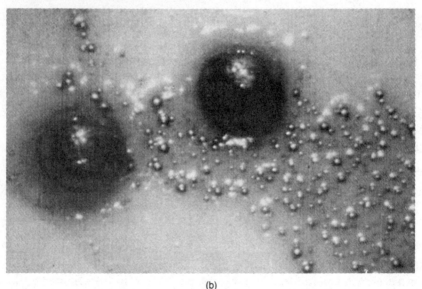

(b)

Figure 7.20   Solder ball incoming inspection test sample on an alumina substrate. (*a*) Solder paste reflowed with no solder ball formations; (*b*) solder paste reflowed with extensive solder ball formation.

1. Deposit a 0.500-in.-diameter by 0.008-in. minimum thick pattern of solder paste at the center of an alumina or PCB substrate.

2. Bake out the sample at 80 to 100°C for 30 to 45 min or with the same schedule used in production.

3. Reflow the sample on a hot plate or, preferably, by the reflow process used in production.

4. Examine the area around the solidified solder sphere for solder balls and determine its conformance with the specified requirements.

Viscosity. Viscosity can be considered the internal friction resulting when a layer of material (fluid, paste, etc.) is caused to move in relation to another layer. A highly viscous material is one possessing a great deal of internal friction; as a result, it will not pour or spread as easily as a material of lesser viscosity. The major parameters influencing the viscosity of solder paste are the quantity of thixotropic agent used in the flux binder, the percent metal content, particle shape, and temperature. The viscosities of solder pastes used typically range from 200,000 to 800,000 centipoise (cP) depending on the type of application technique used. The viscosity of the solder paste plays an important part in the ease of its application to the PCB and the quality of the print when applied to the PCB during the processes of printing and baking out.

A viscometer is a tool used to measure solder paste viscosity by calculating the force required to rotate a spindle in the solder paste. Differences in solder paste premixing, spindle type, spindle's rotation speed, time-to-read value, and temperature will give different values. Therefore, the viscosity of solder paste must be measured in the same way for values to be related to one another. The viscosity test method described below provides a typical technique:

1. Slowly stir the solder paste in its container so that the flux binder is dispersed evenly throughout the metal powder. Standardize the container such as the one it is received in.

2. Stabilize the temperature of the solder paste and place it on the viscometer (see Fig. 7.21).

3. Insert the recommended (see Fig. 7.22) spindle and take care to avoid trapping air under and around the sides of the spindle.

4. Start the spindle rotation at the recommended speed (see Fig. 7.22) 1 min after the spindle has been inserted into the solder paste.

5. Allow the spindle to rotate for 30 s. Stop the instrument by using the clutch and record the dial reading.

6. Convert the dial reading to viscosity in centipoise, in accordance with Fig. 7.22.

Powder oxidation content. The best solder paste to start with, in one respect, is one which has no or very little metal oxidized powder. During the reflow process, it is possible for solder paste which has oxidized metal powder to coalesce into a solder joint and not produce any solder

Figure 7.21 A 5-ounce jar of solder paste is suspended below a spindle attached to a Brooksfield RVF model viscometer.

Figure 7.22   Recommended Spindles and Rotation Speed for Brookfield RVF Viscometer

Viscosity range, cP*	Spindle	Spindle rotation speed, r/min	Factor†
80,000–160,000	5	2	2000
160,000–200,000	6	4	2500
200,000–400,000	6	2	5000
400,000–800,000	7	4	10,000
800,000–1,000,000	7	2	20,000

*If dial reading is below 20 or above 80 cP, move to the spindle recommended for the next lower or higher viscosity range.

†To obtain the viscosity in centipoise, multiply the dial reading on the 100 scale by the factor for the given spindle and speed.

balls, some of the time. Usually when joint formation occurs, the oxidized metal powder that is not reduced by the flux will rise to the surface of the solder joint, producing a surface with an irregular appearance. When it is necessary to conform to high-reliability military standards, determining the amount of oxidized metal powder may help decrease the chances of using a solder paste that may produce solder balls.

Quantitative measurement of oxide content is possible through the use of an Auger analysis; however, this can be costly and time-consuming. The metal powder oxidation can be rated qualitatively by a simple method as described below:

1. Approximately 10 g of solder paste is placed in a crucible along with enough peanut oil to render the container approximately three-quarters full.

2. The crucible is then placed in an oven at 210°C for 30 min to reflow the solder. During this time the peanut oil extracts the flux from the paste so that it cannot clean the oxide from the metal powder; however, it also prevents additional oxidation during the heating and reflow stages.

3. Upon cooling, the peanut oil is decanted and an appropriate solvent is added to dissolve the remaining oil and flux and to facilitate the removal of the solder.

4. The degree of oxidation is clearly indicated by the amount of oxide coverage on the surface of the metal. Oxides will appear as rough, dull growths on the metal's surface.

5. It is preferred, of course, to have zero percent oxide coverage; however, a maximum of 25 percent coverage should be acceptable for most applications.

7.3.3.2 Solder alloys. The solder alloy, whether used in solder pastes, wire, or bars (wave soldering purposes), will not require incoming inspection composition analysis. However, qualification and in-process solder analysis (for wave soldering and static solder pots) should be performed on a typical monthly basis. Composition analysis determines the amounts of metal impurities along with the amounts of the main alloy metals, which typically consist of tin (Sn) and lead (Pb). The criteria for metal purities in tin-lead–bearing solders are specified in QQ-S-571 and in Fig. 7.23.

Wave soldering and lead tinning are the two processes that will pick up metal impurities simply due to contact with the component leads, PCBs, and fixtures when used. Over time the molten solder will continuously dissolve metals from the assembly being soldered, which will typically consist of copper, gold, iron, and zinc. Once they reach their limits, these metal impurities can cause detrimental soldering effects. Depending on which impurity is at its limit, the solder and solder joints can have such characteristics as grittiness, lack of adhesion, poor flow behavior, and brittle conditions from intermetallic growth.

Solder analysis by most companies is simply subcontracted to the vendor supplying the solder and can be performed by quantitative

Figure 7.23   Solder Contaminant Levels for Various Metals as Specified by QQ-S-571E

Contaminant	Contamination limit, %
Aluminum	0.005
Antimony	0.2–0.5
Arsenic	0.03
Bismuth	0.25
Cadmium	0.005
Copper	0.08
Gold	0.08
Iron	0.02
Silver	0.01
Zinc	0.005
Others	0.08

methods such as atomic adsorption. The frequency of testing by military standards (MIL-STD-454, reg. 5) is once per month per solder pot; however, depending on the amount and type of work being soldered, longer periods of time between analysis could serve just as well.

7.3.3.3 Fluxes.   Often it is left to the flux manufacturer to determine the quality level of the flux provided in solder pastes or in liquid form for wave- and hand-soldering techniques. The lack of inspecting flux prior to its use in manufacturing can be the single factor causing defects on manufactured assemblies. Fluxes, which are further explained in Chap. 8, are available in various types, but all serve the purpose of removing metal oxides and tarnishes from the metal surfaces to be reflowed or soldered along with aiding the flow of solder during its molten stages. The incoming inspection of flux should consist of the following tests:

1. *Water extract resistivity:* This test, as specified in QQ-S-571, provides a measure of the ionic nature of the flux system. Unactivated fluxes such as rosin and mildly activated fluxes (RMAs) are classified as *rosin* fluxes, which have a water extract resistivity of not less than 100,000 $\Omega \cdot$ cm. Fluxes which have a water extract resistivity of less than 100,000 $\Omega \cdot$ cm are considered to be activated and are not acceptable for use on military assemblies due to their potential corrosiveness and higher levels of ionic materials remaining on the PCB after the cleaning process.

2. *Copper mirror test (optional for typical commercial assemblies):* This test, also specified in QQ-S-571, consists of a measurement of the activity of the flux by its effect on a thin layer of copper supported by

a glass-mirror substrate. For a R or RMA flux to conform to QQ-S-571, regardless of its conformance with the water extract resistivity test, it must not be active enough so as to remove any of the copper or the glass-mirror substrate. It is possible for a flux to pass the water extract resistivity but fail in the copper mirror test. This type of flux would be unacceptable for use as a R or RMA type when conforming to QQ-S-571.

3. *Specific gravity:* This is a measure of the flux's density, which is greatly affected by its solid content. The specific gravity of the flux during wave soldering techniques can vary significantly depending on the evaporation rate of its solvents and the amount of product being fluxed. As received from the vendor, this characteristic is usually not a problem; therefore, in-process inspection for this flux characteristic should be sufficient. This can consist of taking samples throughout the day, as necessary, and measuring the specific gravity using a hydrometer. This technique is time-consuming and also requires a temperature calculation adjustment. For high-volume manufacturing, using an on-line automatic flux density control system, as shown in Fig. 7.24, is recommended. This type of system constantly displays the flux's specific gravity and triggers the addition of flux or thinner to the flux pot once the specific gravity goes beyond its preset limits.

4. *Flux formulation:* This test, which should be optional, is a precaution that may be taken to ensure that the manufacturer's flux has a consistency of ingredients. An infrared spectrophotometer can be used for this analysis to determine if any new readings or peaks are observed that are different from the standard reading.

5. *Color:* Flux color change or a difference in color, as received, can be an indication of chemical instability or breakdown due to its exposure to light or heat and sometimes simply due to its shelf life. This test, which can be performed visually with a trained eye, is the simplest method of testing; however, a colorimeter is the best method for this analysis. Do not take this test lightly, especially when synthetic and synthetic-rosin-based fluxes are used.

7.3.3.4 Adhesives. Adhesives are used for many purposes on SMAs such as for thermal transfer, component vibration dampening, and component bonding to the PCB. In any event, we are concerned with the adhesion of the adhesive to that to which it is bonded and whether or not any adhesion ability is lost due to the quality of the adhesive or how it is mixed prior to release to manufacturing. Beyond this discussion the user should also be aware of other adhesive characteristics which may have an effect on circuit performance such as corrosivity,

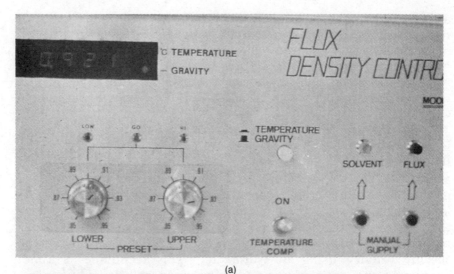

(a)

(b)

Figure 7.24 Automatic flux density (specific gravity) control system for wave soldering machines. (*a*) Control panel; (*b*) flux pot mechanisms. (*Courtesy Sensby Corporation.*)

volume resistivity, ionic impurities, outgassing, and its dielectric constant.

The amount of adhesion an adhesive has to its matting surfaces is called its bonding strength. This characteristic should be measured at incoming inspection when the material is received and periodically during its specified shelf life, especially when the adhesive is used for bonding SMDs during dual-wave soldering. The bonding strength of the adhesive holding a SMD to the PCB can be determined as specified in MIL-STD-883. However, for the purposes of convenience and cost, a simple shear test can be performed by bonding an SMD such as a chip capacitor to the same material to which it is being bonded on the actual assembly (such as solder mask or bare substrate). The amount of adhesive applied and the curing schedule should be the same used in manufacturing. Then a hand-held force gauge or one of the more automated types is used to apply enough force to shear the SMD away from the bonded adhesive.

Performing a test such as this may indeed indicate it is time to terminate the adhesive that has seen extended shelf life or has been affected by moisture absorption or exposure to ultraviolet radiation. Remember, components that are bonded for wave-soldering purposes do not fall off in the wave solder due to defective components.

7.3.3.5 Solvents. Incoming inspection of the cleaning solvent used to remove flux residues from the SMAs is necessary from the standpoint of cleaning efficiency and safety. Most solvents that are used are either azeotropic mixtures or solvent blends and contain a base solvent, alcohol(s), stabilizer(s), and, in some cases, acid acceptor(s). Each ingredient plays an important role in the performance of the solvent, and each must be maintained in its proper percentage. In most cases, the manufacturer of the solvent will provide a gas chromatograph (GC) analysis for each lot of solvent received. Typically this service is done free of charge when asked for. The GC provides the type of each ingredient and its percentage. Normally, the GC as received for a solvent will be within the manufacturer's specified conditions.

During the solvent's use in the manufacturing process is where these ingredients, regardless of whether or not an azeotropic mixture or solvent blend is used, can change beyond their specified operation percentages (by weight). Once this occurs, the solvent can become flammable, corrosive, and a poor cleaner depending on which type of solvent brand is used. For this reason, incoming inspection plus in-process inspection are vital in maintaining an effective cleaning process. In-process inspection of the solvent can simply be done by taking samples of solvent from each tank in the cleaning system and performing a GC on each sample. If a solvent's GC analysis tends to show a significant shift in its composition or an ingredient out of compliance, consulting with the solvent manufacturer should be an immediate task.

References

1. JEDEC Solid State Products Engineering Council, "Editorial Committee Letter Ballot JC-11. 1-84-72," February 15, 1984.
2. William Smith, "The Effect of Lead Coplanarity on PLCC Solder Joint," *Surface Mount Technology*, June 1986.
3. Michael Arcidy, "Maximizing Chip Resistor Performance and Reliability for Surface Mount Applications," *Evaluation Engineering*, September 1986.
4. Clyde F. Coombs, Jr., *Printed Circuits Handbook*, Third Edition, McGraw-Hill Book Company, New York, 1988.
5. C. A. Harper, *Handbook of Electronic Packaging*, McGraw-Hill Book Company, New York, 1969.

Bibliography

Allen, B. M., "Testing For Solderability," *Connection Technology*, May 1986.
Becker, Gert, "Solderability Testing of Surface-Mount Devices," *Electronic Packaging and Production*, October 1987.
Bonner, J. K., "Flux Fingerprinted by Ionic Content," *Circuits Manufacturing*, January 1987.
Davy, J. Gordon, and Skold, Randy, "Solderability for Receiving Inspection," *Circuits Manufacturing*, February 1985.
Manko, H. H., "Soldering Fluxes—Past and Present," *Welding Journal*, March 1973.
Roos-Kozel, B. "Reliable Solder Paste Quality Control," *Brazing and Soldering*, no. 10, Spring 1986.
Weinberger, L., and Audette, D., "Energies of Activation and Rates of Solder Wetting with Activated Fluxes," *Insulation Circuits*, July 1980.
Wild, R. N., "Component Lead Solderability vs. Artificial Steam Ageing II," *Brazing and Soldering*, no. 13, Autumn 1987.

Manufacturing Materials and Application Techniques

8.1 Manufacturing Materials

8.1.1 Solder pastes

Within the last 17 years an evolution has been taking place with solder paste materials. Originally developed to satisfy industrial and automotive applications, solder pastes have gone through a transition to become a highly refined electronic material. This transition from industrial to electronic applications has required a significant development effort since for many surface mount designs, solder paste technology has replaced the conventional method of applying solder to an assembly, which is the well-known wave soldering process.

Solder pastes available today are refined to the extent that they compare with standard hybrid thick film materials in printing characteristics and reproducibility and at the same time can satisfy military requirements for compatibility with high-reliability electronic assemblies. Solder paste is a homogeneous, prealloyed suspension of solder powder in a thixotropic flux-containing material, hereon called a flux system. The design of the flux system with the solder powder along with its interaction before and during its use is the major concern when using solder pastes. The rheology behavior, the solder powder characteristics, viscosity, and metal load, and the flux system characteristics, all a major concern, have an effect on the performance of the solder paste during the application, bakeout, and reflow steps of the process. To help our understanding of this complex material, let us review the principle of rheology and follow this with the basic characteristics of solder pastes which will aid in our selection of the various solder paste formulations and lead to success in their proper application.

8.1.1.1 Rheology. Rheology is simply the science of a material's flow properties under stress.[1] The most commonly used term is viscosity, which is technically the ratio of shear stress to shear rate. Consider a rheological model composed of a rectangular body of exceedingly thin layers of viscous material stacked on top of each other, similar to a deck of playing cards. If a force is applied causing it to move as in when solder paste is printed, it tends to pull the the viscous material layer underneath along with it, and so on through the material. The velocity gradient through the body of viscous material is defined as shear rate. If shear stress is defined in dynes per centimeter (dyn/cm) and the shear rate is stated in reciprocal seconds (s^{-1}), the viscosity (cm^2) is the ratio of shear stress to shear rate, expressed in poise (P) or centipoise (cP).

Various types of material flow are defined as newtonian, plastic, pseudoplastic, dilatant, and thixotropic, as illustrated in Fig. 8.1. For so-called ideal or newtonian liquids such as honey, the viscosity does not alter with changing shear rate. Most printing substances, particularly solder pastes, are nonnewtonian. Materials belonging to the plastic group are perhaps the clearest indicators of the difference between newtonian and nonnewtonian materials. They also illustrate how deceiving the term "thick" can be when applied to viscosity. The common household condiment, ketchup, is a plastic material, while honey, as mentioned, has a newtonian behavior. When emptying a ketchup and a honey bottle at the same time, the honey will flow freely while the ketchup will not even drip. Normally we would say that the ketchup is thicker or more viscous than the honey, whereas exactly the opposite is true. If we hit the bottom of the ketchup bottle a few times or apply shear stress, it will start to flow and will empty much faster than the honey. Plastic flow, like the ketchup, is characterized by a "yield" point at which the applied stress overcomes the

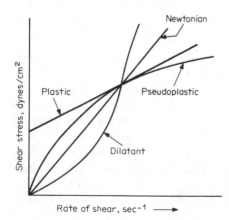

Figure 8.1 The various types of flow characteristics.

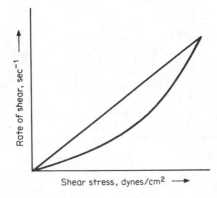

Figure 8.2 A typical flow curve for a thixotropic material such as solder paste demonstrating a hysteresis effect, in that the viscosity at any particular rate of shear will depend on the amount of previous shearing it has undergone.

strength of the inner structure. From then on, the plastic material exhibits a newtonian flow characteristic. Pseudoplastic materials do not have a yield point; thus at the application of stress, flow starts immediately. As the shear rate increases, viscosity decreases in a nonlinear fashion. The pseudoplastic changes may or may not be reversible, but when they are, the changes are instantaneous. Synthetic resins and high polymers are usually in the group of materials which exhibit dilatant flow. Commonly, heavily pigmented materials exhibit a nonlinear change in viscosity as the shear rate increases to a point where they may become a plastic or even a solid material. Thixotropic materials such as solder pastes behave essentially the same as pseudoplastics with one important difference. While the apparent viscosity decreases under increased shear rate in both groups, only thixotropic materials need a finite time to return to their original viscosity. When the apparent viscosity of a thixotropic material is plotted, as shown in Fig. 8.2, the shear rate versus shear stress curves do not coincide during the decreasing and increasing viscosity phases. The size of area within the hysteresis loop is an indication of the material's thixotropic property. Another way to observe the thixotropic behavior is to hold the shear rate constant. Pseudoplastic materials will then exhibit a constant viscosity. Thixotropic materials, on the other hand, will keep on decreasing their viscosity as time goes on until an equilibrium is reached, as illustrated in Fig. 8.3.

From the printer's point of view, the thixotropic behavior of solder pastes is a mixed blessing. Because of the time required to reach equilibrium, thixotropic materials will hold an image's edge better than a pseudoplastic material will. By the same token, they will take a longer time to flow out, which may be detrimental to fast production. Finally, the fact that their viscosity continues to decrease under constant shear rate, such as provided by a screen at a given squeegee speed, means that the solder paste has a lower viscosity on a large

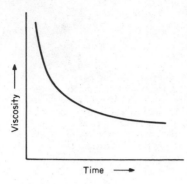

Figure 8.3   Solder paste viscosity decreases under a constant shear rate over a period of time.

screen, at the end of a squeegee stroke, than it had at the beginning. In addition to flooding the screen with solder paste, this action may mean that heavier deposits are produced at the end of a squeegee stroke than at its beginning.

8.1.1.2   Solder paste rheology during printing.   Typical solder paste flow characteristics, as shown in Fig. 8.4, occur during its application by printing techniques. Part of the complexity of solder paste rheology is due to the dispersion of solder powder in a viscous flux system. These dispersions are heterogeneous mixtures with the dispersed phase (solder powder) significantly larger than the media molecules (flux system). The affinity between the flux system and the solder powder occurs primarily on the surface of the individual solder particles. The quality of this dispersion can be determined by a physical characterization of the flow behavior.

Deflocculation is a characteristic of good dispersions, whereas flocculation (chaining effects of particles) is a characteristic of poor dis-

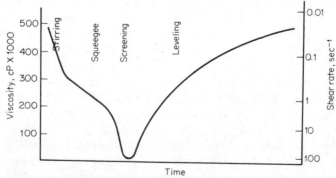

Figure 8.4   Typical solder paste flow characteristics occurring during its application by screen printing techniques.

persion. In a closely packed material such as solder paste, there is only enough flux system to fully separate the solder powder. If low shear is applied, minimum viscosity resistance or change occurs because adequate time exists for the solder particles to relocate without coming into intermittent contact with others. If a high shear is suddenly applied, adjacent solder particles can penetrate the flux system between solder particles, resulting in a solid-to-solid contact. Major viscosity resistance or poor printing characteristics occur due to a lack of lubrication by the viscous flux system. Also, the more irregular the solder particle shape, the greater for this process to occur under sudden high stresses, making it more difficult to print solder paste uniformly.

8.1.1.3 Viscosity. The entire topic of viscosity is ill-defined by both users and manufacturers. Many times, a viscosity value is provided without identifying the type of viscometer, model, or spindle, and without the rate per minute (rpm) of the spindle. As shown in Fig. 8.5, the viscosity value of solder paste varies greatly depending on the rpm, even if all other conditions, including temperature, are held constant. A typical viscosity test condition for solder paste is given in Sec. 7.3.3.1; however, when comparing or using solder pastes as they relate to viscosity, make sure its measurement is performed in the same way, otherwise you may be comparing apples one day and oranges the next.

The major parameters influencing solder paste viscosity are the percentage of metal loading (percent metal content), powder particle size,

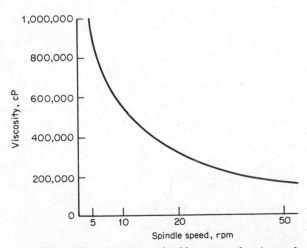

Figure 8.5 Viscosity curve of solder paste showing a decrease as shear rate increases. In this case, shear is produced by the rotating spindle on a viscometer.

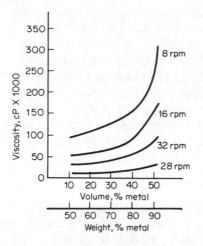

Figure 8.6 The effect of increasing metal loading on solder paste viscosity for four different testing shear rates. (*Courtesy Alpha Metals, Inc.*)

temperature, quantity of flux, and lubricating nature of the thixotropic agents. Figure 8.6 illustrates how metal loading affects solder paste viscosity at four different shear rates (spindle rpm's). In all cases, the increase in metal loading increases the viscosity of the solder paste, if the flux formulation is held constant. However, if more lubricating agent is added, the viscosity can be reduced at a higher metal load. The effect of powder particle size on viscosity of the solder paste with identical metal loading and flux system is illustrated in Fig. 8.7. This shows that as powder particle size increases, the viscosity decreases, and with a decrease in size, the viscosity increases. From this occurrence, it could be stated that as powder particle size

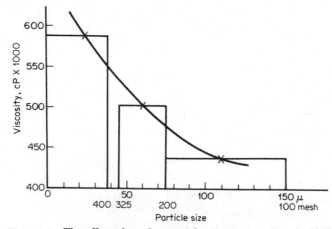

Figure 8.7 The effect of powder particle size on viscosity of solder paste with identical percent metal load and flux system. (*Courtesy Alpha Metals, Inc.*)

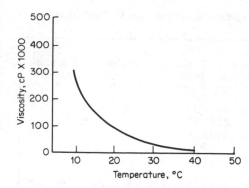

Figure 8.8 Variation in solder paste viscosity for a small range of ambient temperatures.

decreases, an increase in shear force occurs due to the additional surface area afforded by the smaller particles. A relationship between solder paste viscosity and temperature change is illustrated in Fig. 8.8. As expected, an increase in temperature decreases the viscosity, whereas a decrease in temperature increases viscosity. To maintain a controlled measurement of solder paste viscosity and controlled process conditions, it will be important to maintain a controlled temperature since a small temperature change of only ±4°F can vary the viscosity by at least 100,000 cP. A variation of this much viscosity change, plus the effect the application technique such as printing has on viscosity change, can significantly alter the quality of the printed solder paste image.

Now that we understand rheology and the viscosity of solder pastes, the issue that may come to mind is to what is the best viscosity when using a solder paste. The best viscosity of a solder paste depends on the type of application technique and the density and thickness of the solder paste to be applied to the PCB. Stencil printing, which uses an etched metal mask, requires solder pastes with higher viscosities and metal loading than the other application techniques such as screen printing and syringe dispensing. Since stencil printing is typically used as a contact printing method, higher viscosities and metal loading will help prevent the solder paste from flowing under the metal mask, thereby producing a better image quality or definition. Unlike stencil printing, screen printing requires lower viscosities and metal loading since the solder paste has to be pushed through openings in the screen which are partially blocked by a fabric mesh such as stainless steel. Figure 8.9 illustrates typical solder paste viscosities associated with a percent metal load and type of application.

8.1.1.4 Metal loading. With such large disparities between the density of the metal powder and flux medium, the difference between the

Figure 8.9   Percent Metal Loading and Typical Viscosities Used for Rosin-Based Solder Pastes

Type of solder paste application	Typical viscosity, cP at 75°F (5-rpm, no. 7 spindle, Brooksfield RVF)	Percent metal loading (by weight)
Stencil printing	550,000–750,000	90
Screen printing	450,000–550,000	90
Screen printing	550,000–675,000	88
Syringe dispensing*	400,000–600,000	85
Syringe dispensing†	350,000–450,000	80
Pin transfer	200,000–300,000	75

Note: In printing techniques where solder paste thickness exceeds approximately 0.008 in., use middle to higher range viscosities to prevent slumping and poor image definition.
*Large-diameter needle.
†Small-diameter needle.

percentage by weight and volume percentage of the metal constituent can be appreciable. Figure 8.10 illustrates the relationship for a Sn60-Pb40 alloy of 200-mesh average particle size in three different flux systems. Considerations such as the space available for the dispersed solder paste, the ability to remain on a restricted area, and the size and thickness of the solder joint required, all contribute to the choice of metal loading. The usual metal load range is from 75 and 90 percent by weight, although with the versatility that the method of manufacture of these materials offers, almost any formulation is possible. It is important to realize, however, that metal loading as a parameter

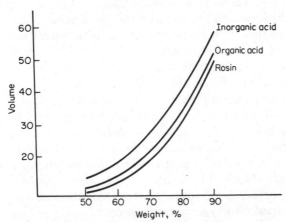

Figure 8.10   The relationship between percent weight of metal and percent volume of metal for 200-mesh solder paste in three different flux-activated systems. (*Courtesy Alpha Metals, Inc.*)

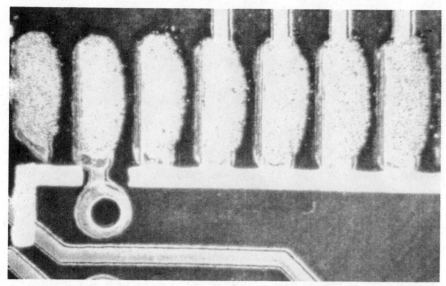

Figure 8.11 Solder paste "slump" during bakeout with 85 percent metal loading, 200 mesh and 550,000 cP.

should not be used alone, because identical metal loadings in flux systems of different activities can lead to significantly different results.

Increasing the metal loading can help to control internal solder joint voids and joint volume and promote reduced cleaning. However, the major effect of increasing the metal loading is the reduction of solder paste "slumping" during bakeout and elevated temperatures prior to reflow. Solder paste slump, as shown in Fig. 8.11, is the collapsing of a deposited layer of solder paste during elevated temperature. More metal loading, which requires greater amounts of thixotropic agents to keep the greater weight of metal powder suspended, and higher viscosities will reduce this slump and prevent solder bridging.

8.1.1.5 Solder powders. Solder powder, the main ingredient of the solder paste formulation, largely controls the performance characteristics of the paste system. The solder powder size, shape, and distribution are the parameters critical to the solder paste's performance. Solder powders are made by using a shot tower in which molten alloy metal is allowed to flow through an orifice into a chamber of controlled inert gas. The flow of the metal stream is disturbed by jet nozzles blowing inert gas under very controlled and proprietary conditions. When the process is well controlled, the spheroidal and uniform-sized solder particles are formed as shown in Fig. 8.12. "Tear drops," "bent bananas," and nonspheroidal solder particles are formed when

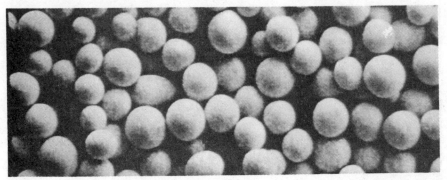

Figure 8.12 Spheroid-shaped solder particles preferred for good solder pastes. (*Courtesy Alpha Metals, Inc.*)

the cooling conditions are not adequate and the powder is "blown" while a portion of the bulk is still liquid.

The majority of the solder pastes available utilize spheroidal type solder powders. This tends to reduce the surface area per given volume, thereby decreasing the available area for surface oxidation. The spheroidal solder particle shape is preferred since it is more consistent and allows for greater reproducibility in terms of the rheological properties of the solder paste. Furthermore, the irregularly shaped solder particles tend to become lodged in the screen mesh, causing erratic printing. Also, irregularly shaped solder particles, which have more surface area, tend to oxidize more frequently, leading to greater chances of producing solder balling.

Solder paste particle size and distribution generally falls into a −200/+325 mesh with no more than 10 percent of the particles falling outside these limits. The −325 mesh distribution is generally used where fine line definition is required for printing solder paste for SMDs with lead centerlines spaces less than 0.040 in.

Particle size and shape analysis can be performed using a scanning electron microscope (SEM), flotation, air classification, or Coulter counter techniques.

8.1.1.6 Flux systems. Flux systems used in solder pastes are flux-type pastes or vehicle materials that, when combined with the metal powder, form a homogeneous mixture. It is the flux system that produces most of the thixotropic rheology in the solder paste. Typically, there are four "components" in flux system formulations which vary in both type and percentages among manufacturers. These consist of the following constituents:

Solvents or alcohols (caritol, glycols, or terpinols)

Rosin or resins (synthetic)

Activators: amines (nonionic, halogenated); amine hydrohalides, organic acids

Thixotropic agents

Solvents combine with the solids in the flux or paste. Preferably the solvent has low moisture absorption along with the appropriate boiling point and vapor pressure depending on whether a short or long drying solder paste is required. Viscosity additives, including rosin and resins, are employed to improve the application characteristics of the solder paste. Activators are suspended in the flux system and are used to remove oxides on the solder particles and the lands to be soldered to. Typically, either RMA (mildly activated rosin) or RA (activated rosin) flux pastes are used with solder pastes. Thixotropes are added to adjust the viscosity for either screen, stencil, syringe, or pin transfer disposition. Thixotropes are also used to help suspend the metal powder, where more are added as the percentage of metal loading increases.

8.1.2 Solder alloys

Solder is a fusible alloy usually containing two to three elemental metals along with several types of impure metals with a melting point below 800°F (425°C). Solder's purpose is simply to join two or more metal surfaces and, in doing so, act as a metallurgical bridge between the surfaces. For solder to reliably connect two joining metal surfaces, it must be able to wet to them. With soldering electronic assemblies, solder is used to join metals such as copper and nickel. With both of these metals, when solder wets them, layers of intermetallics form, which contribute to the final wetting conditions necessary for these materials to join. If these intermetallics do not form, solder wetting to these metals will not occur (with some exceptions). This can be caused by a layer of metal oxides on the surfaces of these metals, which will not allow solder wetting. Metal oxides will occur on such metal surfaces; therefore, fluxes are used to reduce the oxides, allowing the solder to make intermittent contact with the metal surface.

Solder wetting to a metal surface can be defined by the so-called dihedral angle. The dihedral angle is simply the degree of angle (σ) formed between the solder and the surface it is connecting to. This relationship is illustrated in Fig. 8.13 between the dihedral angle and the degree of solder wetting.

There are many types of solder alloy compositions with various characteristics, as shown in Fig. 8.14. The tin (Sn)-lead (Pb) alloys are the most common for electronic applications because they have good strength and wettability, but they are not recommended for soldering to silver, silver alloys, and gold due to the rapid development of brittle

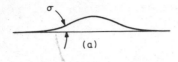

(a)

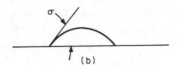

(b)

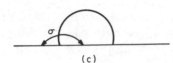

(c)

Figure 8.13 Solder wetting variations: (a) Total wetting ($\sigma = 0°$);
(b) partial wetting ($180° < \sigma > 0$);
(c) total nonwetting ($\sigma = 180°$).

intermetallics (leaching). Lead-indium (In) alloys have a good bonding
strengths in thermal cycling and show reduced leaching against pre-
cious metals. Tin-silver (Ag) alloys have superior wetting and high
strength; however, silver migration is possible. Tin-lead-silver alloys
have minimum silver scavenging and improved creep resistance.

These solder alloys either consist of eutectic, near-eutectic, or
noneutectic compositions. Eutectic solder compositions consist of those
solder alloys that have a single melting point, rather than a melting
range where solid and molten solder particles exist simultaneously.
Near-eutectic solders do not have a sharp melting point but do have a
fairly narrow melting point range with a typical temperature range of
5°F. Noneutectic solders, which are never used for electronic interconnections, include those solder alloy compositions which have a wide
melting point range from as much as 10 to 30°F and possibly more.
Figure 8.15 is a binary-phase diagram for tin-lead solders. In this
case, as the solid and liquid lines meet, the pasty range of the alloy no
longer exists since this composition, Sn63-Pb37, is the eutectic of
these two metals as an alloy.

The two most common metal alloys used in SMT consist of the Sn63-
Pb37 and Sn62-Pb36-Ag2 compositions. The Sn63-Pb37 alloy is used
for both solder pastes and wave solder applications due to its eutectic
characteristic and its satisfactory melting point of 361°F (183°C). The
Sn62-Pb36-Ag2 composition is only used in solder paste or wire solder
when soldering to metal surfaces such as on chip capacitors containing
palladium silver end caps. This alloy, which contains 2 percent silver,
is used to reduce the amount of leaching which occurs between the sil

Figure 8.14   Solder Alloy Compositions

Composition	Tin, %	Lead, %	Antimony, %	Bismuth, max., %	Silver, %	Copper, max., %	Iron, max., %	Zinc, max., %	Aluminum, max., %	Arsenic, max., %	Cadmium, max., %	Total of all others, max., %	Approximate melting range, °C	
													Solidus	Liquidus
Sn96	Remainder	0.10, max.	—	—	3.6–4.4	0.20	—	0.005	—	0.05	0.005	—	221	221
Sn70	69.5–71.5	Remainder	0.20–0.50	0.25	—	0.08	0.02	0.005	0.005	0.03	—	0.08	183	193
Sn63	62.5–63.5	Remainder	0.20–0.50	0.25	—	0.08	0.02	0.005	0.005	0.03	—	0.08	183	183
Sn62	61.5–62.5	Remainder	0.20–0.50	0.25	1.75–2.25	0.08	0.02	0.005	0.005	0.03	—	0.08	179	179
Sn60	59.5–61.5	Remainder	0.20–0.50	0.25	—	0.08	0.02	0.005	0.005	0.03	—	0.08	183	191
Sn50	49.5–51.5	Remainder	0.20–0.50	0.25	—	0.08	0.02	0.005	0.005	0.25	—	0.08	183	216
Sn40	39.5–41.5	Remainder	0.20–0.50	0.25	—	0.08	0.02	0.005	0.005	0.02	—	0.08	183	238
Sn35	34.5–36.5	Remainder	1.6–2.0	0.25	—	0.08	0.02	0.005	0.005	0.02	—	0.08	185	243
Sn30	29.5–31.5	Remainder	1.4–1.8	0.25	—	0.08	0.02	0.005	0.005	0.02	—	0.08	185	250

Figure 8.14 Solder Alloy Compositions (*Continued*)

Composition	Tin, %	Lead, %	Antimony, %	Bismuth, max., %	Silver, %	Copper, max., %	Iron, max., %	Zinc, max., %	Aluminum, max., %	Arsenic, max., %	Cadmium, max., %	Total of all others, max., %	Approximate melting range, °C	
													Solidus	Liquidus
Sn20	19.5–21.5	Remainder	0.80–1.2	0.25	—	0.08	0.02	0.005	0.005	0.02	—	0.08	184	270
Sn10	9.0–11.0	Remainder	0.20, max.	0.03	1.7–2.4	0.08	—	0.005	0.005	0.02	—	0.10	268	290
Sn5	4.5–5.5	Remainder	0.50, max.	0.25	—	0.08	0.02	0.005	0.005	0.02	—	0.08	308	312
Sb5	94.0 min	0.20, max.	4.0–6.0	—	—	0.08	0.08	0.03	0.03	0.05	0.03	0.03	235	240
Pb80	Remainder	78.5–80.5	0.20–0.50	0.25	—	0.08	0.02	0.005	0.005	0.02	—	0.08	183	277
Pb70	Remainder	68.5–70.5	0.20–0.50	0.25	—	0.08	0.02	0.005	0.005	0.02	—	0.08	183	254
Pb65	Remainder	63.5–65.5	0.20–0.50	0.25	—	0.08	0.02	0.005	0.005	0.02	—	0.08	183	246
Ag1.5	0.75–1.25	Remainder	0.40, max.,	0.25	1.3–1.7	0.30	0.02	0.005	0.005	0.02	—	0.08	309	309
Ag2.5	0.25, max.	Remainder	0.40, max.	0.25	2.3–2.7	0.30	0.02	0.005	0.005	0.02	—	0.03	304	304
Ag5.5	0.25, max.	Remainder	0.40, max.	0.25	5.0–6.0	0.30	0.02	0.005	0.005	0.02	—	0.03	304	380

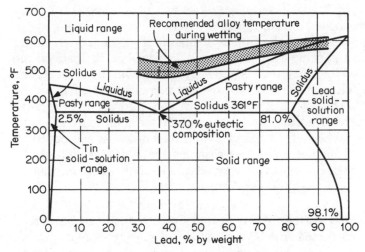

Figure 8.15 Binary phase diagram for tin-lead solders.

ver in the end cap and the tin in the solder. This, alloy, however, actually has very little effect in accomplishing a reduction in leaching since the dwell times at liquid temperatures are long enough to accelerate the leaching beyond any effects the silver can have in the solder alloy.

8.1.3 Fluxes

Fluxes are chemically and physically active substances or formulations which enable wetting of the metal surface to be soldered by removing the oxide or other surface films from the metal surface and the solder itself. Fluxes also protect the metal surfaces from reoxidation during soldering and reduce the molten solder's surface tension, promoting solder spread and flow.

Flux is applied to the surfaces to be soldered either by conventional wave soldering methods such as foaming and waving or as a flux paste. During heating, most fluxes reach their activation temperature in order to clean the metal surfaces while the solvents carrying the flux are evaporated. Solvent evaporation is necessary prior to reflow temperatures since in sufficient quantities, spattering of the molten solder will occur. Whether used in wave soldering or solder paste fluxes, choosing the right solvent or solvent mixture according to their boiling points can make significant quality differences, especially when using solder paste technology. During the molten stages of solder, the flux is displaced from the surface during wetting and from within the molten solder when solder pastes are used. After flux displacement occurs, it still has a major effect on the outcome of the quality of the solder fillet formed, especially in dual-wave soldering applications. In this case, once the flux is displaced after the molten solder wets the lands and component leads or metallizations, it floats on top of the molten solder, affecting its rheology and surface tension. As the percentage of solids in the flux formulation increases, its effect on the molten solder's rheology increases, which can significantly improve the molten solder's ability to make contact with difficult areas to reach on dual-wave-soldered SMAs. When they are designed not to affect circuitry performance, the flux residues are sometimes left on the assembly; sometimes they are removed due to their effect on insulation resistance and possibly corrosivity.

8.1.3.1 Rosin-based fluxes. At room temperature, rosin is a hard, natural material extracted from the oleoresin of pine trees and refined, consisting primarily of abietic and primaric acids and their isomers, some organic fatty acids, and terpene hydrocarbons. Rosin is available as gum, wood, and tall oil, is sometimes chemically modified, and comes in various grades of color. The purest form of rosin by the ASTM designation is called *water white,* or WW; it is the mildest type of flux and is referred to as nonactivated rosin.

Water white rosin itself is usually too mild to remove sufficient amounts of metal oxides for achieving effective wetting of the solder. To improve this condition, activators are added such as alkylamine hydrohalides and organically bound halogen compounds. This has resulted in several different types of rosin-based fluxes with the following designations:

R: rosin, nonactivated

RMA: rosin, mildly activated

RA: rosin, activated

RSA: rosin, superactivated

Mil-F-14256, QQ-S-571, and IPC-SF-818 control the requirements for R and RMA type fluxes, which are the only types of fluxes acceptable for use on high-reliability electronic assemblies. They consist of rosin-based fluxes which can pass the copper mirror test along with having a water extract resistivity of not less than $100,000 \ \Omega \cdot cm$. These two types of rosin fluxes are noncorrosive; the RMA type is most commonly used in liquid form for wave soldering purposes and in paste form for solder pastes. The RA type flux is extensively used in commercial and consumer products and is often left on the assembly in products such as radios, TVs, and telephones. When left on the as-

sembly, the RA fluxes' water extract resistivity usually ranges near
80,000 Ω · cm. The RSA fluxes are used for special applications, typically for cleaning difficult metals to solder, and are certainly not suitable for use on any type of electronic assembly due to their corrosiveness and difficult removal.

Another group of fluxes which contain rosin and are currently un-
defined by either military or commercial specifications are becoming
extremely popular for use in dual-wave soldering. These fluxes consist
predominantly of synthetic resins, which are commonly used in synthetic fluxes, and small amounts of rosin. A variety of these fluxes are
available with a range of activity levels. Although no official designations exist for these fluxes, the author characterizes them in the same
way as pure rosin-based fluxes, as follows:

SR: synthetic resin, rosin; nonactivated

SMAR: synthetic resin, rosin; mildly activated

SAR: synthetic resin, rosin; activated

SSAR: synthetic resin, rosin; superactivated

The basic advantage of these new groups of fluxes is their ability to
be removed more easily than pure rosin-based fluxes. During assembly cooling, rosin-based fluxes become hard solids and tend to be more
difficult to remove as the activity of the flux increases, whereas with
these synthetic resins, removal after wave soldering is much easier,
even with higher activities, due to the fact that the flux residues are
soft and fluid at room temperature.

8.1.3.2 Synthetic fluxes. Synthetic fluxes, developed by Dupont, consist of synthetic resins, solvents, foaming agents, and activators. They
were primarily designed for conventional wave soldering (single
wave) processes, in order to provide the industry with an activated
flux which is capable of achieving cleanliness levels compared to those
reached by RMA fluxes. The other advantage with these fluxes is that
they do not leave behind the so-called insoluble white residues (see
Chap. 11) common with the rosin-based fluxes.

A disadvantage with these fluxes has to do with their use on
dual-wave soldering systems. Dual-wave soldering systems, which
have strong solder "scrubbing" activity, tend to wash off these
fluxes because they have low solid contents, which results in solder
icicles and opens. These defects are simply caused by a lack of flux
on the PCB when it enters the second, molten solder wave. The synthetic resin, rosin-based fluxes have been designed to overcome this
problem.

8.1.3.3 Organic fluxes. Organic fluxes, sometimes called organic acid fluxes, resemble the superactivated rosin fluxes except that they are rosin-free and water-soluble. Their use in SMT is not very promising since they are normally classified as corrosive and must be removed from the assembly. The need to remove these fluxes with water poses a serious concern regarding the ability to remove these flux residues from under the small gaps beneath most SMDs. Although organic fluxes have been extensively used in soldering conventional assemblies with good results, their ability to compete with the current advances in solvent soluable activated fluxes is unlikely.

8.1.4 Adhesives

8.1.4.1 Theory of adhesion. For an adhesive to bond to a material, it must make intimate contact with the material; any type of contamination such as oil and grease, which can be simply a fingerprint, must be removed for wetting of the adhesive to occur. Once the adhesive wets a material, several forces that promote and produce adhesion come into play. These forces consist of both physical and chemical characteristics.

The physical forces which contribute to the role of adhesion are mechanical. The surface area and interlocking cavities of the surface to be bonded to are the critical areas affecting the mechanical parts of adhesion. Once the adhesive makes contact with its bonding surface, it will tend to flow and displace the air within the micropores of the bonding surface. As this occurs, the surface area to which the adhesive is making contact can increase significantly. The larger the surface micropores, the greater the penetration the adhesive will have below the apparent surface layer. As the surface area and penetration increase, the adhesive's mechanical bonding strength increases.

Even to this day, the chemical characteristics of adhesion are still theoretical; however, they can be explained based on the molecular attraction between the adhesive and bonding surface. Each surface or material has a certain affinity for each other during their contact. For example, water will flow and spread more on clean metal than it will on wax. This is basically due to the variations in surface tension which play an important role in wetting. Once wetting occurs, both materials start to become attracted to each other's weak molecular forces, which accounts for the chemical adhesion. These weak molecular forces are called Van der Waal's forces.

8.1.4.2 A word about adhesives and their uses. From a functional standpoint, adhesives are structural, nonstructural, or sealing. Structural adhesives, which are materials of high mechanical strength, are used to permanently bond materials together, sometimes with the function of holding materials together under a load. Nonstructural adhesives are not designed to hold substantial loads but are widely used to hold things in place. Nonstructural adhesives are commonly used to bond SMDs for wave soldering purposes. Sealing or potting adhesives do not have any significant load-bearing capability and are intended for gap-filling or vibration-damping requirements. These adhesives are normally very flexible, whereas the other two types of adhesives are typically rigid.

From a chemical standpoint, adhesives can be classified as either thermosetting, thermoplastic, elastomeric, or alloy. Thermosetting adhesives are materials which cure by chemical reaction to form a cross-linked polymer. Once cured, subsequent heating and softening cannot be used to renew or establish new adhesive bonds; however, reheating near and above the cured adhesive's glass transition temperature can substantially reduce its adhesion strength. Thermosetting adhesives are available as one- or two-part systems. Generally one-part systems require elevated temperatures to cause rapid curing. Two-part systems can rapidly cure at room temperature but require accurate mixing of the resin and catalyst to achieve the proper specified adhesive characteristics and curing. Thermosetting adhesives are the types used for bonding SMDs to the PCB. Some chemical examples of thermoset adhesives are epoxies, cyanoacrylates, acrylics, and polyesters.

Thermoplastic adhesives do not form a cross-linked polymer like thermoset adhesives and, therefore, may resoften any number of times to renew or form new adhesive bonds. Thermoplastic adhesives are one-component systems that harden upon cooling from an elevated temperature or upon the evaporation of a solvent. Some chemical examples of thermoplastic adhesives are polyesters, polyvinyl acetals, acrylics, polyamides, and cellulose acetates.

Elastomer adhesives are materials with large elongation properties. These adhesives may be formulated from synthetic or naturally occurring polymers in solvents, latex cements, or dispensions. Some chemical examples of elastomer adhesives are urethanes, silicones, neoprecores, and natural rubbers.

Alloy adhesives are formulated by combining thermoset, thermoplastic and elastomer adhesives in a manner that will utilize the most useful property of each material. Some chemical examples of alloy adhesives are epoxy-nylons, epoxy polysulfides, and vinyl-phenolics.

8.1.4.3 Adhesives used for SMT.

There are two major types of adhesives that are used for SMT—epoxies and acrylics. Both types of adhesives are commercially available with various characteristics such as bond strength, viscosities, pot lives, curing techniques, and curing schedules. Each type of adhesive can literally be formulated with dozens of characteristics required by the user. This becomes an important issue when manufacturing assembly is limited as to how it cures or applies these adhesives.

Epoxies are the oldest and most used adhesives. Epoxies, as mentioned, are thermoset, structually strong adhesives and are available as liquids, pastes, films, and powders in one- or two-component systems. One-part epoxies require an elevated temperature to induce rapid curing and refrigerated storage to inhibit the very slow polymerization which occurs at room temperatures. These one-part epoxies will cure in conventional ovens with circulating air and like all epoxies can have a wide range of viscosities and various curing requirements depending on temperature and dwell time. Two-part epoxy adhesives are composed of the epoxy resin and a second part known as the catalyst, hardener, or curing agent. These two parts must be mixed together to cause polymerization curing of the epoxy. Once the two components are mixed, depending on the formulation, curing will proceed at room temperature within minutes or hours or will be accelerated at high temperatures. When properly formulated, such adhesives can be used to bond chip SMDs without a special curing process step prior to wave soldering. Care should be taken to verify if sufficient bonding strengths are maintained by the adhesive during wave soldering.

Acrylic adhesives, which are becoming very popular for use in bonding SMDs for dual-wave soldering, are relatively new compared to epoxy adhesives. The acrylic adhesives are based upon the chemistry of methacrylate or substituted methyl methacrylate and are usually cured using a short exposure to ultraviolet radiation and a longer exposure for 1 to 2 min at approximately 150°C using infrared radiation. This typical curing schedule for acrylic adhesives is substantially shorter than for most epoxy adhesives; however, acrylic adhesives cannot be cured at room temperature. This means curing acrylic adhesives requires the appropriate equipment, which can cost in excess of 50,000 U.S. dollars. Epoxies, on the other hand, can be cured in a simple convection oven costing several hundred dollars up to conveyorized convection ovens costing less that 15,000 U.S. dollars. Furthermore, for the best bonding strength results, portions of the acrylic adhesive must go beyond the component body so as not to be shielded from the exposure to the ultraviolet radiation cure step.

8.2 Solder Paste Application Techniques

The application of solder paste to PCBs can be performed by several methods: screen (or stencil) printing, syringe dispensing, or pin transfer techniques. The screen printing method is extensively used for solder paste application, whereas the syringe dispensing and pin transfer techniques are more commonly used for adhesive application. Although these application techniques have been used in various technologies, obtaining a quality image or transfer of the material requires an in-depth understanding of the particular machines, processes, and materials being applied. The basic goal, no matter what application method is used or material applied, is to obtain the same resolution, thickness, and registration to the substrate for as many times as necessary with as few variations in these conditions as possible. The following sections will provide us with the basic fundamentals of these application techniques.

8.2.1 The principles of printing

Printing is an application technique in which all the areas where the printing material is to be located can be treated simultaneously. The two types of printing techniques are off-contact and direct contact printing. The off-contact printing uses a screen which supports an emulsion with openings. The solder paste is flooded over the screen so that it is uniformly spread and fills the openings in the emulsion. The squeegee then pushes down on the screen, causing the screen area along the squeegee to make contact with the PCB. As the squeegee passes over the emulsion openings, it pushes the solder paste through and onto the PCB.

Direct contact printing consists of using an etched metal stencil instead of a screen. The metal stencil, which has etched unblocked openings, is mated in direct contact with the PCB during the entire printing cycle. Flooding of the solder paste is not necessary and, therefore, only the application of the solder paste requires the printing stroke.

8.2.2 Printing machines

There are three types of printing machines: the manual, semiautomatic, and automatic. The manual printing machine, as shown in Fig. 8.16, is the least expensive and consists of an x,y table and a supporting structure which holds the frame. The frame height above the PCB and the alignment are controlled manually by the operator along with a hand-held squeegee blade which the operator uses to spread and print the solder paste.

The semiautomatic printing systems, as shown in Fig. 8.17a and b, consist of the very portable and less expensive desk-top models for low- to moderate-volume workloads and the stand-alone systems primarily used for production manufacturing. Each system works in the

Figure 8.16   Manual printer. (*Courtesy Elite Circuit Equipment.*)

same way, such that the head which supports the screen or stencil frame, the squeegee head, and the down-and-up stroke of the squeegee are pneumatically controlled. The flood bar is either at a preset fixed position or is also controlled pneumatically in its up-and-down stroke.

The automatic printing systems have the same operating technique as the semiautomatic printing systems with the exception that the PCBs are automatically loaded, aligned, and unloaded without the assistance of the operator. Normally, the operator sets up the machine parameters, applies solder paste to the screen or stencil as necessary, and keeps the on-loader magazine filled with PCBs. The off-loading of a PCB printed with solder paste is normally activated by the in-line placement machine as boards loaded with components are finished. Figure 8.18 illustrates an automatic PCB loader connected to an automatic printing system.

8.2.3   Basic functions and mechanisms of printing machines

8.2.3.1   Substrate holding mechanisms.
A printing machine must have a means of holding the PCB and sometimes a PCB assembly in position throughout the printing operation. This can be accomplished by simply relying on the tooling pins for PCB registration and pressure-sensitive tape to hold the PCB corners flat. The best approach is using a vacuum table or platen. Vacuum tables come in various designs; however, there are those that work, of course, and those that do not. A common type of vacuum table, as shown in Fig. 8.19, has numerous

(a)

(b)

Figure 8.17 Semiautomatic printers. (a) Portable, desk-top type (*Courtesy Elite Circuit Equipment*); (b) stand-alone production type (*Courtesy MPM Corporation*).

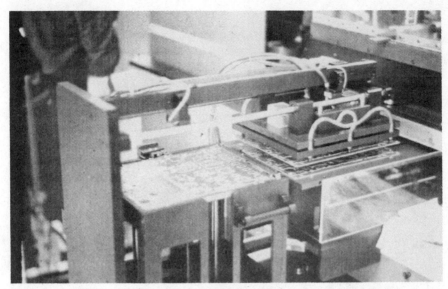

Figure 8.18 PCB board loader connected to an automatic printer.

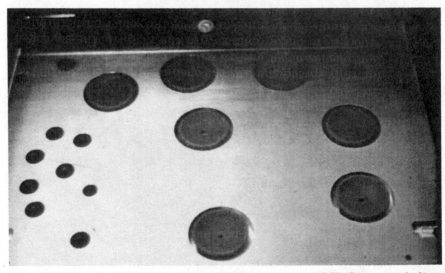

Figure 8.19 Vacuum table with suction cups for maintaining PCB flatness and alignment during printing.

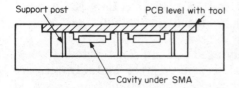

Support post   PCB level with tool

Cavity under SMA

Figure 8.20 Special tooling for holding a double-sided SMA while printing.

rubber vacuum cups that reside below the surface of the table. Once the PCB is located above the vacuum table, the vacuum cups are raised above the table, producing a tightly scaled vacuum on the PCB, and then they are quickly lowered, pulling the PCB firmly flat against the table. During this process, the PCB is also aligned with expanding tungsten carbide pins. When planning on purchasing a printer with a vacuum table, it is strongly recommended to verify if your worst PCB (the one with the most PTHs) can be held flat against the vacuum table during its operation. A good vacuum system will hold flat even the most warped PCBs with many PTHs.

The above type of holding mechanisms which has a flat table can only be used when the bottom of the PCB has no components. When there are components on the bottom side of the PCB, the table design must change. In this case, the table is designed so that it holds and makes contact with only the edges of the PCB and is recessed below the height of the components. Figure 8.20 illustrates this type of table, which can be further designed to pull a vacuum. With or without vacuum, supporting posts underneath the assembly are necessary to prevent any bowing of the board during printing.

8.2.3.2   Alignment.   All initial PCB to screen (or stencil) alignment is done visually by the operator. The best approach to proper alignment is to lower the head so that the screen lies almost flat against the PCB but with a very small space to allow for movement of the screen above the PCB lands. The operator should then look vertically above the screen and move the x, y, and σ controls, which will either move the table or head to a point where satisfactory alignment is achieved. The first few printed PCBs should always be checked for proper solder paste alignment and corrections made when necessary. The printer machines which are the easiest to align are those which have low front profiles so that even short operators can easily look down on the screen and those that have electronically driven heads or tables.

8.2.3.3   The printing frame.   Printing frames are used to support the screen or stencil and to maintain them parallel to the table which supports the PCB. Printing frames are manufactured in an extremely wide range of sizes. For surface mount applications, they are normally standardized in sizes of 12×12 in., 15×15 in., 16×16 in., and 20×20 in. The two basic types produced are called rigid or self-tensioning.

Rigid frames are the most widely used and are constructed of wood, anodized aluminum, stainless steel, or printed steel. Metal frames are commonly used for surface mount applications. Metal frames are, of course, more expensive than wood, but their dimensions are not affected either by changes in relative humidity or by degreasing and emulsion or stencil preparation steps involving water or aqueous chemicals.

Self-tensioning frames are less commonly used and are more expensive than rigid frames. These enable the user to make readjustments of the mesh fabric tension as it relaxes with age or usage. They are not used for stencils.

Frame size and type may be entirely controlled by the type of printing machine used; but for optimum registration and better longevity of the screen and squeegee blade, it is advisable to select frame dimensions which allow a minimum 4-in. clearance around the printed area of the screen. If stencils are used, this clearance can be reduced; however, it would still be good practice to maintain this clearance in case screens are used at some point. This clearance is a factor that should not be taken lightly and should be a major concern when purchasing a printing machine. The smaller the clearance between the squeegee blade, the more wear and tear occurs to the screen and blade, along with an increasing difficulty in obtaining a uniform solder paste print image. This relationship is be illustrated in Fig. 8.21*a* and *b* where proper printing conditions are reached when the clearance between the edges of the frame and squeegee blade is long enough. When this clearance is too short, squeegee blade distortion results at the corners, producing an irregular print in both thickness and resolution.

8.2.3.4 The squeegee assembly. The squeegee assembly is probably the most complicated moving mechanism on the printing machine. It consists of a squeegee blade and holder, speed pressure controls, and a flood bar or blade as shown on the printing system in Fig. 8.22. The squeegee assembly is used to perform the following functions:

- When screens are used, it spreads out an even layer of solder paste across the entire screen area with the flood bar or blade. This printing step, called the flood stroke, is illustrated in Fig. 8.23. Flooding is not necessary on stencils since there is no mesh across the openings and flooding is not used to apply solder paste to PCB.

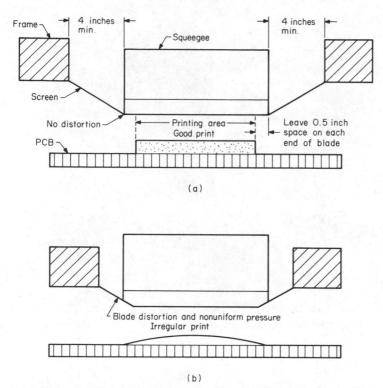

Figure 8.21 Relationships between frame and squeegee as they affect printing quality. (*a*) Proper conditions; (*b*) distorted conditions.

- It holds the screen in momentary contact with the PCB and adapts the screen to the irregularities of the PCB surface.

- It removes the excess paste from the screen, leaving behind the solder paste which fills up the screen or stencil openings. Once the screen and stencil lifts off the PCB, it leaves behind the solder paste. This printing step, called the print stroke, is illustrated in Fig. 8.24.

The squeegee blade is an elastomer, typically made of polyurethane. The blade material must be highly resistant to abrasion and to the solvent used for its cleaning. The hardness of the blade material is an important characteristic which can affect the print quality of the solder paste. Normally the durometer of the blades used for solder paste printing ranges from 60 to 90 (A shore). When screens are used, the lower durometer ranges are used, whereas for stencils the higher ranges are used. The shape of the blade comes in two configurations, rectangular or square. They are typically purchased in foot lengths

Figure 8.22   The squeegee assembly and controls. (*Courtesy MPM Corporation.*)

Figure 8.23   Solder paste is uniformly spread over the screen during the flood stroke.

Figure 8.24 After flooding (for screen printing) the print stroke moves solder paste through the screen and piles the rest of the paste at one end of the screen for the next flood stroke.

and when received are cut to the length of the blade holder. In all cases, the blade must be attached to the blade holder so that it can lie perfectly flat on a flat surface. If the blade is not well connected to the blade holder and does not lie perfectly flat once it is secured, its contact with the screen or stencil will not be uniform in pressure across the length of the blade. This will result in very irregular print characteristics, such as solder paste smudging, poor resolution, and locations having little or no solder paste.

The blade holder is usually a simple clamp or bolt mounting design that allows the extension of the blade beyond its edges. A proper blade holder design allows an ease of blade assembly so that it is not distorted when the blade holder is tightened to secure the blade.

The squeegee speed, or the rate at which it travels in direct contact across the screen or stencil, directly affects the print quality and thickness. If the drive force is too slow, the squeegee cannot move with consistent speed since frictional forces will make it vary. In addition, the speed must be constant across the entire screen or stencil without undue acceleration or deceleration.

Squeegee pressure is one of the most important and misunderstood parameters in printing solder pastes. Excessive squeegee pressure causes unnecessary friction against the screen or stencil. This in turn causes the blade to wear quickly or lose its sharpness and can distort the screen or the stencil quite rapidly. The resulting conditions, possibly in just a few dozen prints, can produce poor quality printing. The best approach in determining the proper squeegee pressure is to start with insufficient pressure and increase gradually just to the point where effective printing results occur.

The flood bar or blade is only used to evenly spread the solder paste over the screen prior to the print stroke. The flooding stroke, which is not necessary when stencils are used, pushes the solder paste partially down into the screen openings. If the flood bar is set too low, it can actually print some of the solder paste or it can make contact with the screen, both of which should be prevented. There are two types of flood bars, a blunt metal blade and a polyurethane blade identical in design with the squeegee blade and blade holder. Most metal blades tend to drag the solder paste instead of shearing it into a thin, smooth surface; however, with proper setup they can be made to perform effective solder paste flooding. The squeegee flood blade, separate from the squeegee (print) blade, offers the best method of flooding and provides for the simplest setup and greatest operating flexibility.

8.2.3.5 Screens and stencils. Screens, like stencils, are used to apply the exact image to be printed and to act as the primary controller of solder paste deposit. The screen can be made of polyester, nylon, or stainless steel mesh fabric. The polyester and nylon fabrics, due to their durability and lower cost, are often used for consumer printing techniques, solder mask, and photoresist applications. Stainless steel fabric is used in the application of solder pastes because of its high dimensional stability, excellent abrasion resistance, and high open area. This fabric can withstand high tensions with minimum extension, hence enabling smaller snap-off distances between the fabric and PCB surface than with polyester or nylon. This characteristic of stainless steel fabric minimizes the solder paste image's distortion and mis-

Figure 8.25   Screen Mesh Data

Mesh count (MC), per centimeter	MC per inch	Mesh opening (MO), cm	Wire diameter (D), cm	% Open area (OA)
23.6	60	0.0309	0.0114	0.533
31.5	80	0.0267	0.0051	0.706
31.5	80	0.0224	0.0094	0.496
41.3	105	0.0166	0.0076	0.469
47.2	120	0.0146	0.0066	0.473
53.1	135	0.0130	0.0058	0.475
57.1	145	0.0119	0.0056	0.464

alignment during the print stroke. Stainless steel fabric comes in many variations according to the mesh count, thread diameter, and opening. Figure 8.25 provides this information for various mesh counts. Typically, solder paste is printed with 60-, 80-, or 105-mesh per inch stainless steel fabric with a common rule being that the solder particle diameter should be a maximum of one-third the size of the mesh opening.[2] The thickness of the solder paste printed onto the PCB is determined by several factors such as mesh size, thread diameter, and emulsion thickness. To determine the solder paste thickness the following equation can be used:

$$PT = (FT \times \%OA) + ET$$

PT = print thickness
FT = fabric thickness
$\%OA$ = percentage open area of the fabric (see Fig. 8.25)
ET = emulsion thickness

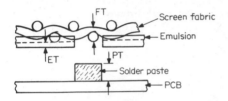

A stencil, which consists of an etched metal foil with mesh-free openings, can be used as a contact printing method for printing solder paste thicknesses greater than what can be achieved with screens. The thickness is determined by the thickness of the metal foil. Stencils can be an all-metal mask or a flexmetal mask. The all-metal mask is connected directly to the frame, whereas the flexmetal mesh is surrounded by either a polyester or stainless steel fabric, which allows for ease of uniform contact to the PCB surface. The single most important limitation of metal stencils is their lack of adaptability to the PCB. Unlike screens, they lack flexibility and the gasketing ability of the polymer emulsion applied to the screen's fabric. Printing over uneven circuitry becomes especially difficult when small, separate solder paste deposits must be made close together.

8.2.3.6 Solder paste printing problems and solutions. Printing is a process as "old as the hills" and has been used to apply "inks" to various surfaces such as bottles, T-shirts, calendars, and hybrids. Typically, these ink deposits are very thin, ranging from 0.002 to 0.0005 in. Solder paste printing, however, requires deposits ranging from 0.004 to 0.030 in. with the most common thicknesses ranging from 0.008 to 0.012 in. This thickness condition along with the complex rheology and chemistry of solder pastes make its application significantly more difficult. In any event, when printing problems do occur with solder pastes, there are ways to overcome them. Let us now review these common printing problems and their potential solutions.

Problem	Cause/Solution
Incomplete print  Solder paste Land (pad) 	1. Clogged screen or partially contacting stencil. 2. Incomplete solder paste flooding. 3. Particle size distribution in solder paste, predominantly consisting of too large particle diameters.
Peaking	Screen off contact distance too large.
Smudging	1. Excessive squeegee pressure or down stroke. 2. Excessive solder paste clinging to bottom of screen or stencil. 3. Viscosity or percentage metal content of solder too low for thickness of print.
Thinner print	1. Poor flooding. 2. Inadequate squeegee pressure or down stroke. 3. Out-of-spec. emulsion or stencil thickness.
Poor edge definition	1. Solder paste viscosity slightly too low. 2. Plating on lands too thick, creating a convex meniscus. 3. Worn-out emulsions.

Problem	Cause/Solution
Irregular print thickness	1. Emulsion layer on screen not uniform in thickness. 2. Screen or stencil not parallel to PCB surface. 3. Irregular flood thickness. 4. Solder paste not uniformly mixed.

8.3 Adhesive Application Techniques

Adhesives which are used for component bonding to the PCB are normally applied by either the syringe dispensing or pin transfer technique. The syringe dispensing method applies a single dot of adhesive per step, whereas the pin transfer technique applies adhesive to all the areas which require adhesive, simultaneously.

Adhesives which work well for syringe dispensing may not work well for pin transfer and vice versa. In most cases, it is the thixotropic behavior of the adhesive that will affect the quality of the adhesive's deposit. When trying out either application technique, a wide variety of adhesives from various manufacturers should be tested for their behavior while they are being applied.

8.3.1 The principles of dispensing

The syringe dispensing and pin transfer technique principles of applying adhesive are relatively the same. When adhesive is applied by either method, it must separate completely from either the needle or pin; otherwise adhesive "stringing" or inaccurate and random amounts of adhesive will be deposited. For proper adhesive separation to occur, there are several conditions that must take place. First, the adhesive must make a solid contact with the PCB surface and wet it. Second, the wetting forces of the adhesive to the PCB surface must be greater than the internal cohesive forces of the adhesive to the needle or pin. In most cases, since adhesives wet more easily to metal surfaces than organic surfaces such as PCBs, the adhesive separation will occur within the adhesive material itself. If the wetting forces of the adhesive to the PCB surface are low, then poor depositions can result. If the adhesive has strong internal cohesive forces, then stringing of the adhesive can occur. Ideal adhesive conditions resulting in quality adhesive dot control would consist of any adhesive which has high wetting forces to the PCB surface which are significantly greater than the internal cohesive forces and its wetting forces to the needle or pin.

Furthermore, the cohesive forces of the adhesive are preferred to be significantly less than the force wetting it has to the needle or pin.

8.3.2 Adhesive application machines

Figure 8.26 illustrates a computer numerically controlled syringe dispensing machine. Machines such as this one are preprogrammed for adhesive dispensing to PCB locations along with a preset volume of adhesive to be deposited. Air under pressure is simply applied to the syringe containing the adhesive which forces the adhesive out the needle. The amount of adhesive applied is affected by the air pressure, the air pressure dwell at each location, the internal diameter of the needle, and the rheology of the adhesive. Some syringe dispensing machines come equipped with a syringe temperature control jacket which maintains the adhesive's temperature within a few degrees. This will help assist in controlling a consistent paste viscosity and aid in more uniform adhesive dispensing.

Figure 8.27 illustrates an automatic adhesive pin transfer machine. From inside the machine, the adhesive is spread out over a plate with a consistent thickness. After this occurs a plate holding pins of various diameters is lowered to make contact with the adhesive. Upon contact, the adhesive wets the pins and is removed. When the plate moves up and over the PCB, the adhesive is applied to the PCB as the plate is

Figure 8.26 Computer numerically controlled syringe dispensing system. (*Courtesy Automation Unlimited, an ICON Company.*)

Figure 8.27 Semiautomatic pin transfer machine. (*Courtesy Heller Industries, Inc.*)

lowered and the pins make contact with the PCB surface. The same cycle is repeated for each PCB.

Instead of using pins, a recent development uses a magnesium alloy plate which is chemically etched to create the pins. This method works extremely well, with infinite amounts of adhesive-applied shapes possible. The cost of producing the plate, which requires artwork generation, is far lower than the tooling required for pin transfer.

References

1. Stephen Ruback," Surface Mount Technology: Part II Deposition of Solder," *SITE*, 1986.
2. Carmen Capillo, "The Assembly of Leadless Chip Carriers to Polyimide Printed Circuit Boards," *Nepcon Technical Proceedings*, February 1983.

Bibliography

DeCarlo, John, "SMT Machine Selection: New Flexibility in Modular Pick and Place Designs," *Surface Mount Technology*, February 1988.

Martel, Michael, "Solder Paste, Part I," *SITE*, May 1986.

Meeks, Stephen, and Roelse, James, "SMT Research and Production at IBM," *Surface Mount Technology*, February 1988.

Ruback, Stephen, "Surface-Mount Technology, Part III: Solder Printing Equipment," *SITE*, December 1987.

Socolowski, Norbert, "Solder Pastes for Hybrid Applications," *Microelectronic Manufacturing and Testing*, October 1982.

SMD Placement, Rework, and Workmanship

9.1 SMD Placement Techniques

The rapid development of SMT and its new component packaging designs have spurred an evolution in automatic SMD placement machines. The very basic approach of automatically placing SMDs is common to most SMD placement machines. This consists of picking up the component from a feeder via vacuum spindle and head, squaring the component for PCB land alignment and placing the component on the PCB lands upon release of the vacuum. This would be considered completing one pick-and-placement cycle for a particular component.

There are essentially three types of automatic SMD placement machines including sequential placement, in-line placement, and mass transfer. Sequential placement machines perform the typical pick-and-placement component cycle except that in some cases the PCB is in a fixed location and the placement head travels back and forth for each function. These stand-alone placement machines (see Fig. 9.1) normally can place SMD from 1500 to 6000 components per hour from various types of component packaging such as tape and reels, tubes, bulk, and trays. The other type of sequential placement systems normally called "chip shooters," have a rotating pick-and-placement head much like an inverted gattling gun with many small working heads that rotate at very fast rates. The PCB is located underneath the rotating heads and moves for proper component positioning and placement. Depending on the manufacturer, these machines are capable of placing SMDs at a rate of 6000 to 16,000 per hour. The fastest component placement rates are achieved only with chip resistors, chip capacitors, and SOT 23s from 8- or 12-mm tape and reels. Figure 9.2 il-

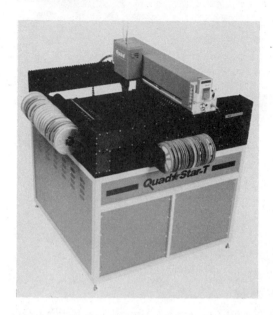

Figure 9.1 A stand-alone sequential SMD placement machine. (*Courtesy Quad Systems Corporation.*)

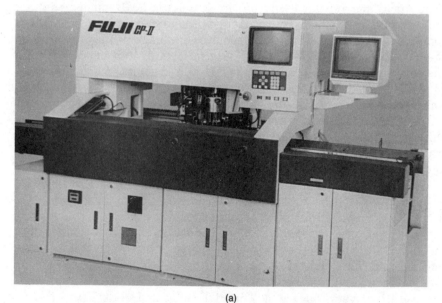

(a)

Figure 9.2   A stand-alone high-speed SMD placement machine. (*a*) Fuji Model CP-II; (*b*) rotating component placement head. (*Courtesy Fuji Machine Mfg. Co., Ltd.*)

(b)

Figure 9.2   (*Continued*)

lustrates a high-speed component placement machine along with its rotating component placement head.

In-line component placement machines consist of any type of sequential machine which is mated via a conveyor to at least one other type of placement machine. This is illustrated in Fig. 9.3 where two of the same type of stand-alone sequential placement machines are connected by a conveyor. This in-line approach to SMD placement decreases the flexibility of manufacturing assemblies, especially when various assembly types exist. The advantages, however, consist of faster assembly throughputs and "modulation" benefits where, if one system breaks down, there are still others to perform the work. This approach is becoming more popular than having one high-speed component placement system, which can cost three to four times the amount than a slower, stand-alone machine capable of in-line manufacturing.

The conveyors (see Fig. 9.4) used for in-line placement machines or those capable of in-line manufacturing usually consist of chain or belt, and carry the PCB on its edges only. Sensors along the conveyor are used to control the movement of the PCB through each machine or placement station.

Mass component placement machines are used where very high volume manufacturing assembly is required. These systems use the same pick-and-place technique as found on sequential placement machines

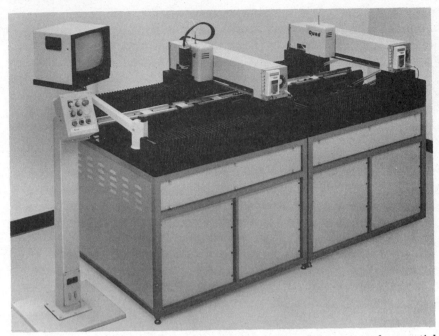

Figure 9.3   An in-line SMD placement machine consisting of two connected sequential machines. (*Courtesy Quad Systems Corporation.*)

Figure 9.4   A conveyor processing an SMA through an in-line assembly machine. (*Courtesy Quad Systems Corporation.*)

except that multiple heads are used which simultaneously pick up components and place them in different PCB locations. Currently, there are only three manufacturers of component mass placement machines: Universal Instruments, Philips, and Panasonic.

Universal's mass placement machine (RhyMAS) is more of a high-volume, multihead system than a "true" mass placement machine. Its competitive approach to high placement rates is added flexibility. This system can be modularly expanded up to 70 placement heads, averaging a total of 30,000 chips per hour.

The Philips mass placement machine (MCM 111) has 32 heads per module and can place 32 components simultaneously in 2.5 s. This comes with significant restrictions on PCB design and component layout. Up to 12 modules can be used together with in-line substrate transfer, resulting in a component placement rate of 552,000 chips per hour.

Panasonic (Matsushita Electric) supplies a mass placement machine, called the Panasert MM, capable of placing 576,000 chips per hour. This machine consists of up to four in-line mass placement stations, each placing 200 components in 5 s, for a maximum total of 800 components in 5 s.

9.2   Major Placement Machine Features

9.2.1   Pick-and-place heads

Pick-and-place heads are designed to pick up a single component, hold it securely during transport, and then release it upon positioning on the PCB. There are presently four methods used for picking and placing SMDs. Three of these involve ensuring that the component is properly centered prior to placement.

The simplest and most inexpensive head design is a vacuum spindle or nozzle. Although adequate for many applications, it does not compensate for component off-center pickup caused by oversized tapes pockets, vibratory feeder pickup locations, or slippage due to rapid transfer.

Two methods have been devised to provide a means of centering the component under the vacuum spindle. The first involves mechanical centering jaws active during component transfer, as shown in Fig. 9.5. The jaws close and center the component; they also open during a pause just prior to component placement. This ensures that the jaws do not contact the solder paste or surrounding placed components. In most cases, different size jaws are required for a range of component sizes. Some machines are capable of automatically changing different size jaws when required.

The second method of component centering involves using separate

Figure 9.5 Component centering jaws or chucks located on the placement head. (*Courtesy Quad Systems Corporation.*)

centering jaws not integral to the placement head. The vacuum spindle first places the component on a centering nest for mechanical alignment. It then picks up the component again for placement on the PCB. The disadvantage of this method is that it requires a double pick-and-place operation for each part.

Self-centering mechanical grippers, another, less common approach, allow simultaneous pickup and automatic centering without the need for vacuum. In this case, a pair of tweezer-type jaws hold the component while centering it along one axis. Certain PCB design restrictions must accommodate this use of the technique.

9.2.2 Component feeders

Feeders, which present the SMD to the machine's placement head, consist of tape-and-reel, vibratory, magazine, and matrix tray types. Tape-and-reel feeders (see Fig. 9.6) are commonly used for chip and discrete SMDs. However, their use for surface mount ICs is becoming more popular due to the larger number which can be put on each reel as compared to magazine or stick feeders. Tape-and-reel feeders, for all types of SMDs, dominate the type of feeders used due to their higher reliability.

Vibratory feeders, as shown in Fig. 9.7, are used for bulk chip resis-

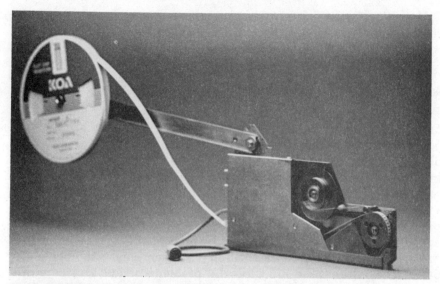

Figure 9.6   Tape-and-reel feeder.

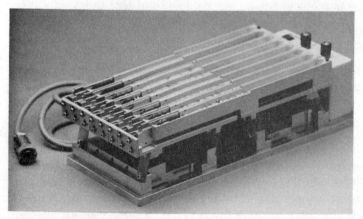

Figure 9.7   Vibratory feeder.

tors, capacitors, and SOT 23s. They literally vibrate bulk components so that they eventually line up in one single row which exposes a component at the front end of the feeder for pickup. Vibratory feeders require that the components have fairly tight dimensional tolerance, otherwise component jamming results.

Magazine feeders are used when parts come in tubes or sticks. Only one tube per feeder is allowed, which may require frequent changes of tubes. In some cases, the components must be removed from the tube and manually fed down a ski-slope vibratory chute.

Matrix trays are simply plastic embossed trays which carry one component per cavity. They are widely used for gull-winged leaded components.

9.2.3 Component Z-axis control

The component Z-axis control has the capability to program the force and depth at which the component placement head will place the component onto the surface of the PCB. This feature, which is not on all placement machines, can control the depth the component is inserted into the solder paste. The accuracy of the depth of the SMD into the solder pastes, especially for chip components, can play a vital role in soldering yields as they relate to solder opens, component tombstoning, and solder balling.

Figure 9.8 illustrates four different component Z-axis relationships consisting of ideal, top contact, tilted placement, and improper solder paste. The latter three are typically problematic and can be eliminated with proper component Z-axis programming. The preferred or "ideal" component placement, as it relates to its Z-axis, is within approximately 50 percent of the solder paste's total thickness.

9.2.4 Vision systems

There are two categories of machine vision systems: vision assistance and vision inspection. Vision assistance analysis consists of pattern error correction and component-to-land matching. Vision inspection analysis consists of component leads, land patterns, solder and adhesive, component presence-orientation, and solder joint inspection.

In general, vision systems (see Fig. 9.9) allow a computer to acquire a digital description of an object (leads, lands, solder paste, etc.) through an intermediate component such as a solid-state camera. This object description comes in a number of formats, including color or gray scale, and can have various fields of views and resolutions.

Field of view and resolution are inversely proportional to each other with machine and vision. With a system having a 256×256 pixel image and a 1-in. field of view, each pixel represents one 1/256 of an inch (approximately 0.004 in.). If the field of view is reduced to a quarter of an inch, the pixel resolution approaches 0.001 in. Machine vision systems can resolve geometries down to ± 1 pixel, which yields accuracies in the 0.001- to 0.002-in. range with no great difficulty. Although recent advances have taken place, the major drawback with vision systems is their inability to digitize complex relationships within short periods of time.

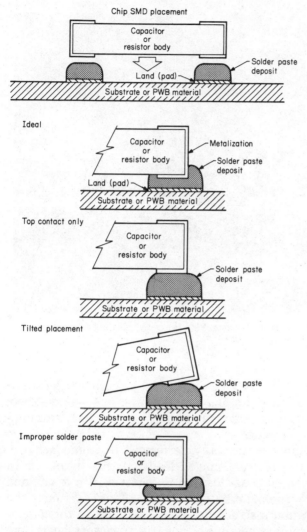

Figure 9.8 Z-axis chip placement criteria. (*Courtesy Dynapert-HTC, Concord, Mass.*)

9.3 Rework of SMAs

Rework can be defined as restoring any unacceptable condition, such as a solder joint, to an acceptable condition in conformance with the specified requirements. Repair is not the same operation as rework and can be defined as any operation used to restore the electrical or mechanical functionality of a damaged or defective condition not in conformance with the specified requirements.

Figure 9.9   A vision system with camera mounted to the placement head and CRT monitor. (*Courtesy Quad Systems Corporation.*)

Rework of SMAs, to some, may be considered a challenging experience. It is, however, a much simpler process than one would expect and, in fact, is much simpler than reworking conventional through-hole assemblies or components.

Rework on an assembly is normally performed for component removal due to nonfunctionality, damaged leads, or misalignment. In order to perform rework such as this and to perform it in a safe and effective way, the proper operator tools are necessary. Without the proper tools and operation instructions, severe damage to the assembly and even the operator is possible. Prior to any rework operation, the operator should have extensive training on experimental assemblies for both the use of the tool and its safety. The tools commercially available provide component rework techniques via convection or conduction methods. The convection method, which is considered the easiest and most effective, simply applies hot air to the solder joints to be reflowed. The conductive method utilizes metal contact tools such as a solder iron.

There are two types of convection rework tools: handheld portable and the stationary module types. The handheld types, as shown in Fig. 9.10, are lightweight and extremely easy to use. These tools uti-

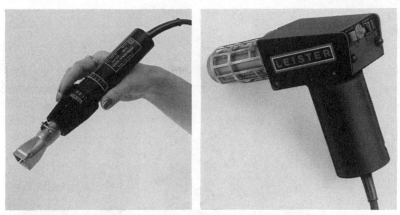

Figure 9.10   Handheld hot air rework tools. (*Courtesy Brain R. White Company, Inc.*)

lize special nozzles designed for each type of SMD being reworked. The nozzles, as shown in Fig. 9.11, are aligned to accurately control and direct the hot air flow to the lands and component leads, preventing reflow of other adjacent component lead solder joints. The component is typically picked up with tweezers or pushed off the land pattern with the hot air tool. The stationary module hot air rework tool, as shown in Fig. 9.12, operates with the same principles; however, it can semiautomatically apply the hot air nozzle over the component and also mechanically remove it via a vacuum nozzle. These rework modules, however, are four to six times more expensive than the handheld types, usually in the range of 9000 to 12,000 U.S. dollars. Another type of stationary hot air repair station, as shown in Fig. 9.13, holds the assembly between two fixed hot air nozzles, one above for reflow and one below for preheating the assembly prior to reflow soldering the component to be reworked.

The conduction solder rework tools consist of the handheld or stationary module types similar to those described for the convection rework tools. These rework tools, however, make an actual mechanical contact with all the leaded sides of the component. The method of reflow is simply by a hot bar, heated by an internal resistive element. These two types of systems, as shown in Fig. 9.14, can be difficult to use effectively since a complete uniform contact between the hot bar and all the solder joints or leads must be accomplished simultaneously. However, with sufficient practice and skill, fast, inexpensive assembly rework can be accomplished. Care must be taken so that all the solder joints have been reflowed before component removal is attempted; otherwise lifting of the lands is inevitable.

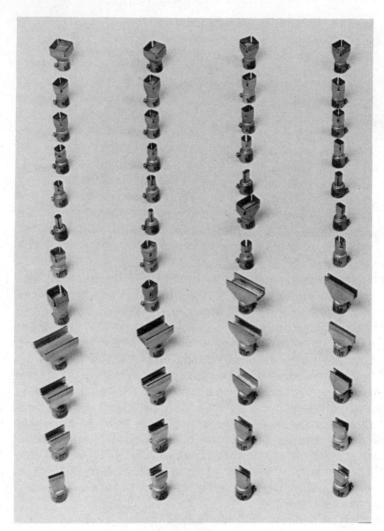

Figure 9.11   Various nozzle designs for hot air handheld rework tools.
(*Courtesy Brian R. White Company, Inc.*)

Figure 9.12   Stationary module hot air rework tool which can semiautomatically remove SMDs. (*Courtesy Air-Vac Engineering Company, Inc.*)

Figure 9.13   Stationary module hot air rework tool with assembly preheating capability. (*Courtesy Nu-Concept Systems, Inc.*)

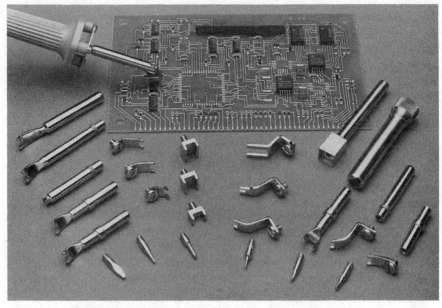

(a)

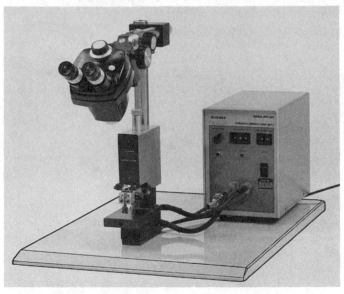

(b)

Figure 9.14   Conduction solder rework tools. (*a*) Handheld (*Courtesy Hexacon Electric Company*); (*b*) stationary module. (*Courtesy Hughes Aircraft Company.*)

9.4   Component Alignment and Solder Joint Workmanship Criteria

Component alignment and solder joint workmanship criteria for SMAs are necessary in order for us to measure or judge the quality of the work performed. What is acceptable or unacceptable should be related to the level of reliability required. In most cases, military and aerospace electronics require higher standards of workmanship than commercial applications. In any case, the higher the workmanship required, whether for a component, a PCB, or an assembly, does not necessarily mean a higher cost to manufacture.

In considering general workmanship criteria for component alignment and solder joints, the reader must realize that correct land geometries are a major factor affecting both of these workmanship conditions. Chapters 5 and 6 provide information on PCB surface mount design, because these design standards may vary according to design restrictions, the following workmanship criteria may also vary. Furthermore, it is strongly recommended that when workmanship criteria are specified, prior review and acceptance of it by manufacturing should take place.

Workmanship: SMD alignment	Chip components

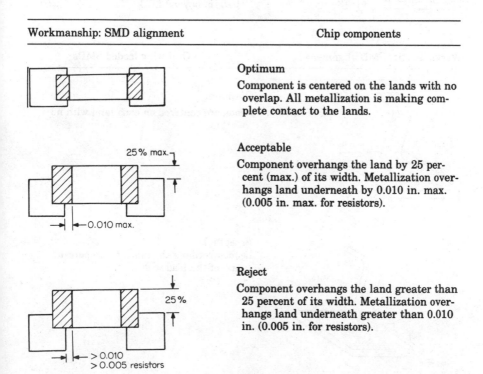

Optimum

Component is centered on the lands with no overlap. All metallization is making complete contact to the lands.

Acceptable

Component overhangs the land by 25 percent (max.) of its width. Metallization overhangs land underneath by 0.010 in. max. (0.005 in. max. for resistors).

Reject

Component overhangs the land greater than 25 percent of its width. Metallization overhangs land underneath greater than 0.010 in. (0.005 in. for resistors).

Workmanship: SMD alignment   Leadless chip carriers (LCC)

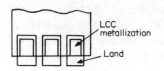

Optimum

LCC metallizations are centered on land pattern with no overhang. Land extends 0.045 in. min. beyond LCC.

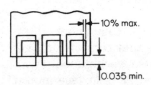

Acceptable

LCC metallizations overhang the lands by 10 percent max. of the width of the metallizations. Land extends 0.035 in. min. beyond LCC.

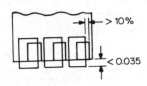

Reject

LCC metallizations overhang the lands greater than 10 percent of the width of the metallizations. Land extends less than 0.035 in. beyond LCC.

Workmanship: SMD alignment   Gull-wing leaded SMDs

Optimum

Leads are centered on each land with no overlap.

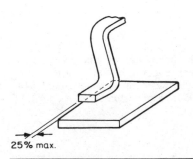

Acceptable

Leads overhang the lands by 25 percent max. of the lead width.

Workmanship: SMD alignment   Gull-wing leaded SMDs

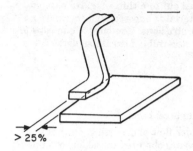

Reject

Leads overhang the lands greater than 25 percent of the lead width

> 25%

Workmanship: SMD alignment   J-leaded SMDs

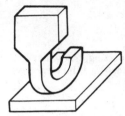

Optimum

Leads are centered on each land with no overhang.

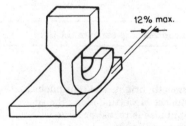

Acceptable

Leads overhang the lands by 12 percent max. of the smallest dimension of the lead width.

12% max.

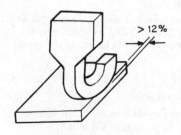

Reject

Leads overhang the lands greater than 12 percent of the smallest dimension of the lead width.

> 12%

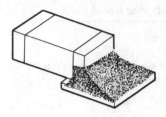

Optimum

Solder is smooth, bright, continuous, and feathered out to a thin edge. No bare end cap material is exposed, and there are no sharp protrusions. Contours of end caps are clearly discernible beneath solder.

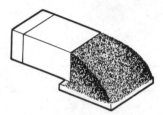

Maximum acceptable

Good solder flow and wetting. Outlines of end caps are obscured by solder, but solder height does not exceed the thickness (height) of the end cap.

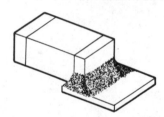

Minimum acceptable

Good solder flow and wetting, but solder fillet coverage is only one-third the thickness of the end cap.

Optimum

Solder is smooth, bright, and continuous and shows evidence of wetting along the edges, but the joint size is robust or large with nearly convex filleting.

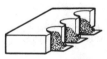

Acceptable

Good solder flow and wetting, but a concave fillet has formed. Fillet height is greater than 50 percent of the height of the metallization.

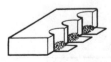

Minimum acceptable

Good solder flow and wetting, but fillet is no less than 50 percent of the height of the metallization.

Optimum

Solder is smooth, bright, continuous, and feathered out to a thin edge. No bare lead material is exposed, and there are no sharp protrusions. A solder fillet behind the heel is evident. Contour of lead is clearly discernible beneath solder.

Maximum acceptable

Good solder flow and wetting, but outline of lead nearly obscured by solder. A solder fillet behind the heel has reached a maximum of one-third the distance to second lead bend.

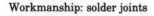

Minimum acceptable

Good solder flow and wetting, but solder fillet only as wide and thick as lead. No fillet behind heel exists.

Optimum

Solder is smooth, bright, continuous, and feathered out to a thin edge. Contour of lead is clearly discernible beneath solder.

Maximum acceptable

Good solder flow and wetting, but outline of lead obscured by solder. Solder behind heel has reached a maximum of 50 percent of the distance to the second lead bend.

| Workmanship: solder joints | Gull wing (small SOIC type SMDs) |

Minimum acceptable

Good solder flow and wetting, but solder fillet only as wide and thick as lead. Solder fillet behind heel is discernible.

| Workmanship: solder joints | J-leaded SMDs |

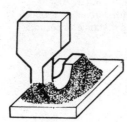

Optimum

Solder is smooth, bright, and continuous and shows evidence of wetting but is robust in size.

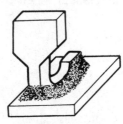

Acceptable

Good solder flow and wetting, but lead contour is slightly discernible.

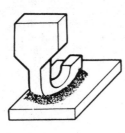

Minimum acceptable

Good solder flow and wetting, but solder fillet is only slightly wider than the lead and solder heel filleting is evident.

Bibliography

Amick, Christopher G., "Close Doesn't Count," *Circuits Manufacturing,* September 1986.

Crosby, Phillip, *Quality Is Free: The Art of Making Quality Certain,* McGraw-Hill Book Company, New York, 1979.

Hall, James W., "Vapor Phase Soldering: A Review of Process Relationships for Successful Surface Mounting," *Nepcon West Proceedings of the Technical Program,* Cahners Exposition Group, Des Plaines, Ill., February 25, 1986.

Markstein, Howard W., "Pick and Place Swings Toward Surface Mount," *Electronic Packaging and Production,* January 1984.

McCraley, Michael T., "Selecting Robotics for Electronic Assembly," *Electronics,* July 1985.

Mead, Donald C., "Machine Vision in SMT," *Assembly Engineering,* August 1987.

Peck, Robert, "Developing Manufacturing Systems for Robotic Insertion Technology," *Electronics,* January 1986.

Rua, Ray, "Automatic Component Placement: Machines for Through Hole and SMT," European Institute of Printed Circuits (EIPC) Conference, December 1984.

Stach, Steve, "SMT Rework and Repair for the 1990s," *The International Society for Hybrid Microelectronics,* vol. 9, no. 1, 1986.

Stewart, James, "Machine Vision Robotics and Printed Circuits," *Northcon 85 Proceedings,* October 22, 1985.

Soldering and Cleaning

Introduction

*Reflow soldering of surface mount assemblies is far more
complex than the soldering technology used for conventional
plated through-hole assemblies. It requires an understanding
of metallurgy, chemistry, fluoroinert fluids, fluxes, component
materials, adhesives, solder pastes, and the various reflow
techniques. All of these disciplines affect the ultimate
reliability and defect level of the assemblies.*

*The central issue in reflow technology is the control of the
reflow process itself. Depending on the type of reflow process,
the particular controls necessary may be dramatically
different. The objective of these controls, however, is not. All
reflow processes should be run in a manner that yields zero
defects. To reach this goal requires that every ingredient of
the reflow process—machines, materials, process
parameters—is carefully evaluated, closely monitored, and in
many cases, significantly modified.*

*The most common first question asked about the various
SMT reflow methods is, What is the best one? A simple
answer is that it depends on the type of SMA, the level of
reliability required, and the particular design parameters
used. Each equipment manufacturer will contend that its
machine offers the best process for most SMAs. This type of
advice is simplistic at best and should be ignored by users
interested in high-quality manufacturing. Unlike
conventional PTH assemblies, one reflow method is not best
suited to all, or even most, surface mount assemblies.*

*Choosing the appropriate reflow method, therefore, will
materially affect the assembly yield, reliability, and total cost
of a particular product. This decision can be a complex one,*

and sometimes the design of the PCB will dictate which reflow process will be used.

Cleaning is normally one of the final assembly processes; therefore, it is one of the last processes which can affect the quality and reliability of the assembly. The word "cleaning," in many instances, may be misleading in that a "very" clean assembly for one application may be a "very" contaminated assembly in another application. For an example, consumer products such as calculators, radios, and cathode-ray tube (CRT) assemblies are often soldered with active fluxes which may have a water extract resistivity of 80,000 Ω · cm and which are never removed from the assemblies or cleaned. Why? Well, because the way the assemblies are used, the flux is not considered a contaminant simply because it does not reduce the performance or reliability below what the assembly is expected to have. However, if the water extract resistivity of the flux is lower, let us say, 50,000 Ω · cm or if the operating or storage temperatures are consistently high, the flux's chemical and electrical effects on the assembly could cause reduced performance or life expectancy; therefore, the flux is now a contaminant and no longer a cosmetic condition. It is therefore necessary to define acceptable assembly cleanliness levels for polar, nonpolar, and particulate matter contaminants. In all cases, acceptable cleaning levels are performed when most or all undesirable matter has been removed from the assembly to an extent that does not impair the circuit performance in the environment and during the expected lifetime it should experience.

What are acceptable cleanliness levels is therefore the major concern and it differs depending on the application and circuitry performance requirements. In all cases, the cleaning process should be tested for its efficiency by performing cleanliness analysis of the assemblies. Furthermore, the added difficulties of cleaning SMAs in very sophisticated military and aerospace hardware, which, by the way, have been successfully cleaned since the 1960s, have caused major controversies. Some of these controversies concern the classification of contaminates, what types of solvents are best, what types of cleaning techniques are best, and how assemblies are to be adequately measured for cleanliness.

These issues and others are discussed in detail in Part 4. The major intent of this section is to provide the reader with a spectrum of working knowledge in order to make the wisest decisions as to what solvents, machines, and processes are best for the product being manufactured and not what is best for the supplier of these related machines and materials.

Soldering Techniques and Equipment

10.1 Vapor Phase Reflow

The vapor phase reflow technique, sometimes called condensation soldering, was developed by Western Electric Company in 1973 and patented by Chu, Mullendorf, and Wenger in 1975. It is now used worldwide and is considered to be one of the most efficient and uniform heat transfer methods available.

When compared with other reflow methods, vapor phase offers the most speed, consistency, and process control when soldering complex assemblies. Assemblies especially suited for vapor phase reflow are those with large quantities of ICs, those with metal cores within the PCB, multiple assemblies in panel form, and multilayer boards. The key characteristic of all of these examples is that the thermal mass is significant, needing an efficient heat transfer method.

Due to its effective heat transfer efficiencies, vapor phase reflow is becoming the preferred process over infrared and hot oil for reflowing tin-lead platings on PCBs during fabrication. For multilayer boards, it is essentially the only quality fusing (reflow) process available to the fabricator other than the messy process of hot oil dipping.

The vapor phase mass soldering technique transfers heat to the product to be reflowed via condensation. Simply stated, vapor phase reflow starts with a boiling fluoroinert fluid that produces a saturated vapor zone at a temperature equal to that of the boiling fluid. Once a saturated vapor zone exists, it displaces most of the air within the reflow zone of the machine, effectively creating an oxygen-free environment. This oxygen-free environment is very important for quality surface mount soldering, since a major cause of solderability problems is the accelerated oxidation of PCBs, components, and soldering paste that occurs at elevated temperatures if oxygen is present.

When the relatively cool assembly enters the saturated vapor zone, the vapor condenses on all the exposed surfaces, imparting the latent heat of condensation to the board, components, and solder paste. Because of the ubiquity of the condensation on all surfaces, vapor phase heat transfer is relatively insensitive to the physical layout or geometry of the assembly. Unfortunately, the efficiency of the thermal transfer at this stage is so great that poor quality components or boards may fail due to the temperature excursion.

Once the assembly is wetted, a continuous phase transformation process has begun. The condensed liquid drains to the bottom of the machine, only to be vaporized and recondense on the assembly, repeatedly imparting the latent heat of the condensing saturated vapor. In a relatively short period of time, everything in the saturated zone reaches a temperature close to, or at, the boiling point of the fluid.

Although this process is thought to achieve uniform reflow of all surfaces and components comprising the assembly, this statement is not entirely correct. As observed with assemblies with components of widely varying mass, the low mass components reach solder reflow temperatures faster than the high mass ones. This phenomenon is not solely controlled by mass but is also affected by the ratio of surface area to mass and the thermal conductivity of the surface material. Fortunately, the additional dwell time necessary to reflow the more massive components is usually only seconds longer than that for the smaller components. As we discuss later in the chapter, this issue is more of a concern for other, less-efficient heat transfer methods such as infrared and hot air.

Because of the constant temperature environment created by the boiling fluid, vapor phase reflow has the attractive attribute of limiting the maximum temperature that any portion of the assembly can see. This maximum can be adjusted by changing fluids or mixtures of fluids, but it is beyond the control of the machine operator, limiting the variation that can be introduced into the process from this source.

10.2 Perfluorinated Fluids

The key to the application of vapor phase reflow to soldering is the availability of a suitable processing fluid. The choice of a fluid for vapor phase soldering is severely limited by the process requirements. First of all, the liquid must have a boiling point high enough to melt the solder but low enough to prevent damage to the assembly. In addition, it must be thermally and chemically stable, compatible with all assembly compounds, free of conductive and corrosive residues, and significantly heavier than air so that it can be easily confined within the system. Ideally, it should also be nonflammable, have low toxicity, and cost little to synthesize. The only fluids known to meet all of these

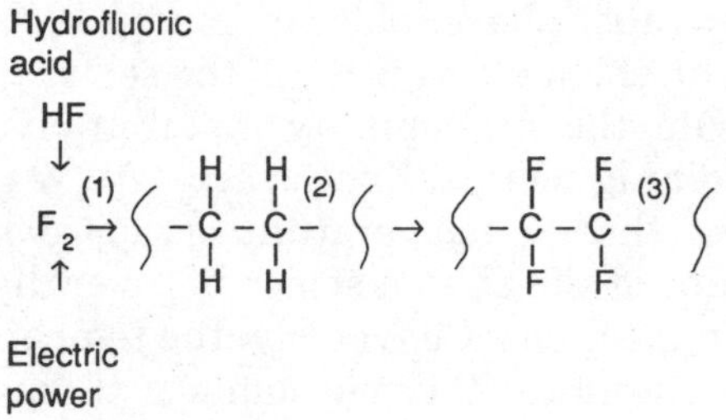

Hydrofluoric
acid

HF
↓
F_2 → $\Big\langle$ $\underset{\substack{| \\ H}}{\overset{\substack{H \\ |}}{-C-}}$ $\underset{\substack{| \\ H}}{\overset{\substack{H \\ |}}{C-}}$ $\Big\rangle$ → $\Big\langle$ $\underset{\substack{| \\ F}}{\overset{\substack{F \\ |}}{-C-}}$ $\underset{\substack{| \\ F}}{\overset{\substack{F \\ |}}{C-}}$ $\Big\rangle$
↑
(1)   (2)   (3)

Electric
power

Figure 10.1 A simplified version of synthesizing a perfluorinated hydrocarbon.

(1) Elemental fluorine is procluded.
(2) Fluorine is reacted with a selected hydrocarbon.
(3) A perfluorinated hydrocarbon is procluded.

requirements (with the exception of low cost) are perfluorinated liquids.

Perfluorinated liquids are members of a family of completely fluorinated organic compounds that have the unique combination of properties mentioned above. The most stable fluids are derived from common organic compounds by the replacement of all carbon-bound hydrogen atoms with fluorine atoms. A common method used to manufacture perfluorinated liquids is by electrochemical fluorination. Although this process does not yield the most stable perfluorinated liquids, which the direct fluorination method does, the process electrolyzes the organic compound in liquid hydrogen fluoride. Figure 10.1 shows a simplified description of the fluorination process.

When the fluorination is complete, the most stable products contain no hydrogen and can have a molecular structure that consists of straight-chain to highly branched or cyclic carbon atoms. In all cases, the molecules are extremely nonpolar and have low solvent action. Their nonpolar character leads to many unusual physical properties, such as low heat of vaporization, low surface tension, and low boiling point in relation to the high molecular weight.

10.2.1 Types of perfluorinated fluids

There are many different types of commercially available perfluorinated liquids, most of which are not intended for reflow soldering purposes. These other fluids are normally used for component gross leak testing, thermal shock, heat transfer, and dielectric coolants. The major reason behind this is simply their boiling point. With boiling points ranging from 133°F to as high as 500°F, only a few offer a boiling point which is suitable for soldering most electronic assemblies.

The solder and solder pastes used for electronic applications (such as eutectic Sn63-Pb36) have melting points at or near 363°F. Therefore, the proper perfluorinated liquid for vapor phase soldering must have a boiling point sufficiently higher than the melting point of the solder to be reflowed and consistently create the high-quality metallurgical bond between the solder and the joining metals. Experience with various soldering processes indicates that a temperature of approximately 50°F above the melting point of the chosen solder is generally needed to produce quality solder connections. Conversely, the temperature must be sufficiently low to minimize thermal damage to any part of the assembly.

In the early development phases of using perfluorinated liquids for reflow soldering, the only available liquid was Freon E5, manufactured by DuPont. Freon E5 is a fluorinated pentapolyoxypropylene having the chemical structure shown in Fig. 10.2. This liquid has all

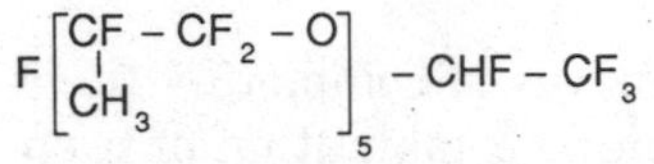

$$F \left[\begin{array}{c} CF - CF_2 - O \\ | \\ CH_3 \end{array} \right]_5 - CHF - CF_3$$

Figure 10.2 Chemical structure of pentapolyoxypropylene (Freon E5).

the necessary physical properties that make it compatible with the process with the exception of its boiling point of 435.6°F. This boiling point is considered excessive when reflowing eutectic or near-eutectic solders. The optimal liquid would have a boiling point in the range of 413 to 420°F, approximately 50°F above the melting point of solder. Certain research, however, shows that slightly higher temperatures may yield better soldering results.

In 1975 vapor phase reflow became a viable mass reflow soldering process when 3M introduced their perfluorinated fluid FC-70, with a boiling point of 419°F. The new liquid, perfluorotrianylamine, had all the necessary physical properties to make it suitable for reliable vapor phase soldering. Its chemical structure contained not only carbon and fluorine atoms, but also nitrogen as shown in Fig. 10.3.

It was initially thought that this so-called fluoroinert liquid was indeed inert or chemically very stable. However, as the vapor phase reflow technique became the most popular soldering method for

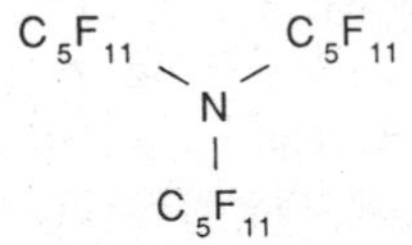

$$C_5F_{11} \diagdown \diagup C_5F_{11}$$
$$N$$
$$|$$
$$C_5F_{11}$$

Figure 10.3 Chemical structure of perfluorotrianylamine.

military and aerospace SMAs, low-level decomposition of FC-70 at its boiling point was observed.

The first evidence of the decomposition of FC-70 was the formation of brown and green residues on the secondary cooling coils within the the vapor phase system. Upon analysis, the green deposits were found to be metal chlorides, and the brown deposits metal fluorides. The metal components of these compounds originated from the stainless steel used in the cooling coils themselves. The green deposits were the result of the decomposition of the secondary fluid, trichlorotri-fluoroethane (FC-113), and the brown deposits the result of the decomposition of FC-70. In each case, the formation of the deposits was rapid, normally occurring within 80 hours of machine operation. The deposit formation period was even shorter when there was superheating due to the use of oversized immersion heaters or the presence of carbonous residues on the immersion heaters.

When the FC-70 decomposition was studied further, it was found that hydrofluoric acid is formed as a by-product of the breakdown of FC-70 into a perfluoroamine, a perfluoroalkane, and a perfluoro-alkene as shown in Fig. 10.4. Once the FC-70 breaks down into a

$$(C_5F_{11})_3N \rightarrow C_5F_{11}N = CF_2 + C_5F_{12} + CF_2 = C(CF_3)_2$$

(FC-70)   (Amine)   (Alkane)   (PFIB)

Figure 10.4 FC-70, during boiling, decomposes into a perfluoroamine, a perfluoroalkane, and a perfluoroalkene (PFIB).

perfluoroalkene (in this case, perfluoroisobutylene, or PFIB), additional reactions occur between the PFIB, water, and alcohol. The reaction of PFIB and water produces hydrofluoric acid directly (Fig. 10.5). The reaction of alcohol with PFIB produces an ether, which can then react with water to form additional hydrofluoric acid (Fig. 10.6).

The formation of a corrosive acid and toxic gases, including PFIB and ether, was unavoidable in the use of the only commercially available perfluorinated liquid. Therefore, the need arose to modify the vapor phase systems to prevent any operator safety hazard and any potential damage to the system itself or the product being soldered.

$$CF_2 = C(CF_3)_2 + H_2O \rightarrow (CH_3)_2 CHCOOH + 2HF$$

(PFIB)   (Hydrofluoric
                                                  acid)

Figure 10.5 PFIB reacts with water to produce hydrofluoric acid during the decomposition of FC-70.

$$CF_2 = C(CF_3)_2 + ROH \rightarrow (CH_3)_2 CH - CF_2 - OR$$

(PFIB)   (Alcohol from   (Ether)
                 fluxes)

$$(CH_3)_2 CH - CF_2 - OR + H_2O \rightarrow (CH_3)_2CHOOR + 2HF$$

(Ether)   (Hydrofluoric
                                                 acid)

Figure 10.6 PFIB reacts with alcohols to produce ether, which then reacts with water to produce hydrofluoric acid.

Although the necessary features were subsequently designed into vapor phase systems, there are still concerns both in operator safety and machine operation when using liquids that can easily break down into unwanted acids and toxic gases.

These concerns regarding the low-level decomposition of the liquid led to the development and commercial availability of more stable perfluorinated liquids. These liquids, with the physical properties listed in Fig. 10.7, include FC-5311 (3M/ICS), LS-215 (Galden), and APF-215 (Air Products and Chemicals). Each one of these liquids provides much more stable conditions with respect to acid formation and PFIB generation during operation than FC-70 and similar liquids, as shown in Fig. 10.8.

10.2.2 Secondary vapor phase fluids

Since the initial commercially available vapor phase systems were of the batch design, the losses of perfluorinated vapors were too high to make the process economically viable. To rectify this situation, Western Electric developed a technique which incorporates a secondary vapor blanket to act as a stable barrier between the expensive perfluorinated vapor and the surrounding air. To create this barrier, the secondary vapor must have a density between that of the perfluorinated vapor and air. In addition, the secondary vapor must be thermally and chemically stable and inert with respect to the perfluorinated vapor and the assemblies. Furthermore, the secondary liquid must have a boiling point lower than the perfluorinated liquid to maintain the condensing scheme necessary for two separate vapor zones. Above all, the cost of the secondary fluid must be low since the loss of this vapor to the environment is relatively high.

The liquid which fits these needs and is used as a secondary vapor is trichlorotrifluoroethane (FC-113). It has a boiling point of 117.6°F, a

Figure 10.7   Physical Properties of Perfluorinated Liquids

	Galden LS = 215*	Galden LS-230	Fluorinert FC-70†	Flutech FC-5311†	Multifluor APF-215‡
Boiling point range, °C	213–219	228–232	213–224	215–219	215–217
Ave. molecular weight	600	650	820	624	630
Heat of vaporization, cal (g)	15	15	16	16	16
Density, 25°C	1.8	1.82	1.94	2.03	2.00
Surface tension, 25°C, dyn/cm	20	20	18	19	21.6
Volume resistivity, 25°C, $\Omega \cdot cm$	1×10^{15}	1×10^{15}	2.3×10^{15}	$> 1 \times 10^{15}$	1×10^{15}
Vapor pressure, 25°C, torr	—	—	< 0.1	< 0.1	0.122
Dielectric constant, 25°C, 1 KH^2	2.1	2.1	1.98	< 2.00	2.00

*Montedison.
†3M.
‡Air Products.

Figure 10.8   Acid and PFIB Generation during the Boiling of Perfluorinated Fluids

	FC-70	FC-5311	LS/215	APF-215
Acid as KOH, mg/g (104 h)	0.100	0.042	0.045	0.041
PFIB	3 ppm	< 0.5 ppb	< 0.5 ppb	< 0.5 ppb

density at 25°C of 97.7 lb/ft^3, and a vapor density of 0.46 lb/ft^3. FC-113 is also miscible in all types of perfluorinated liquids, allowing uniform boiling conditions in the same tank.

The FC-113 vapor is known to be thermally and chemically stable at 117.6°F. However, when it is exposed to the boiling temperature of the perfluorinated liquid (typically 419°F) in the presence of organic matter (such as flux), FC-113 is known to react as shown in Fig. 10.9. The final reaction in this figure produces hydrochloric acid (HCl), which will combine with water and condense on the secondary cooling coils to produce the green deposits mentioned earlier. Fortunately, no

$$F_2ClCCFCl_2 + RH \rightarrow F_2ClCCHFCl + RCl$$
(FC-113)   (Flux)

$$F_2ClCCHFCl \rightarrow F_2CCFCl + HCl$$
                              (Hydrofluoric
                                acid)

Figure 10.9 The low-level decomposition of FC-113 into hydrochloric acid.

harmful gases have been detected during the decomposition of FC-113 under these conditions.

10.2.3 Vapor phase reflow equipment

To fully understand the vapor phase reflow technique when used for soldering SMAs, one must thoroughly understand the equipment design and the correct operation and maintenance. This vapor phase process can yield zero defect soldering without the part movement problems commonly talked about in articles on vapor phase reflow. It can be very cost effective and even less expensive than the wave soldering process.

Initially vapor phase reflow systems existed only as dual vapor batch machines. The pioneering work by Western Electric led to the production of commercial batch systems manufactured by Hybrid Technology Corporation (HTC). However, today several vapor phase equipment manufacturers sell a variety of designs with both batch and conveyorized configurations.

10.2.3.1 Batch vapor phase reflow systems. There are two types of batch vapor phase systems—conventional and thermal mass. Although these two differ in many aspects, they both operate under the same set of basic principles.

All batch systems generate two vapor zones confined in an open-ended, rectangular, stainless steel tank. The primary vapor zone, consisting of the fluoroinert vapor, is located in the bottom of the tank. This is where the assembly is reflowed by the latent heat of condensation transferred during condensation of the vapor. A secondary vapor zone is formed above the primary zone with a less dense, lower boiling point, less expensive fluid such as FC-113. This secondary vapor zone provides several important features which have been critical in allowing the batch system to be effective and efficient. Of major concern is the amount of primary vapor losses to the atmosphere. Primary vapor losses due to convection, diffusion, and dragout are significant enough to make a batch system without a secondary vapor zone, or an inefficient secondary vapor zone, uneconomical. To avoid this,

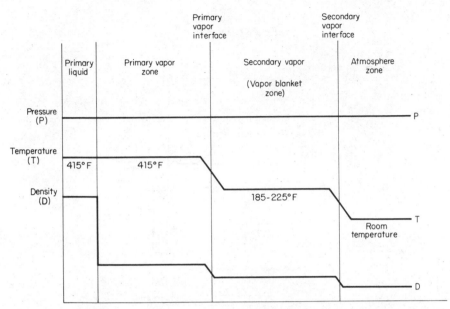

Figure 10.10 Pressure, temperature, and density gradients which exist across both the primary-secondary and secondary-air interfaces in a batch vapor phase reflow system.

all batch systems operate with a secondary vapor zone acting as a cooler vapor blanket or barrier which condenses most of the primary vapor during the machine operation. Furthermore, the secondary vapor zone can act as a preheat zone for the assembly since it typically stabilizes between 180 and 225°F.

In any case, what makes the batch vapor system work is the sharp vapor temperature and vapor density gradients which exist between the three zones along with essentially no pressure gradients, as illustrated in Fig. 10.10.

During system operation, assemblies placed on a elevator platform, which travels vertically through the system or vapor zones, cause a change in both vapor temperature and density gradients that would deviate from the ideal conditions. To minimize changes in the different gradients, which cause an increase in fluid losses and reduced reflow temperatures, these systems are designed to accommodate a more rapid return to ideal gradient conditions by providing more energy or fluid (in thermal mass systems), which leads to producing greater amounts of vapor during the reflow cycle.

10.2.3.2 Conventional vapor phase batch systems. The conventional vapor phase batch system can be best explained by reviewing the cross-sectional illustration in Fig. 10.11. This system's major areas in-

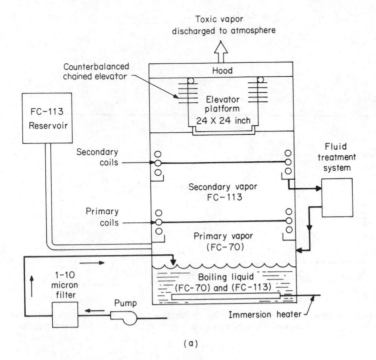

(a)

(b)

Figure 10.11 Conventional batch vapor phase reflow system. (a) Cross-sectional view; (b) external view. (*Courtesy Dynapert-HTC.*)

clude the electrical immersion heaters, condensing coils, fluid treatment system, and flux filteration system. Understanding each one of these areas is essential for operating this process effectively and efficiently.

10.2.3.3 Electrical immersion heaters.

Both vapor zones in this system are created by the continuous boiling of the fluids contained in the bottom of the tank. The fluids are boiled by means of several electrical immersion heaters, which usually have a maximum power to surface area ratio of 20 W/in^2. Higher wattage heaters have been found to produce excessive surface temperatures, which can cause gross fluid decomposition and therefore should be avoided. Furthermore, if higher boiling point fluids are used (greater than 419°F), it may be necessary to use higher wattage heaters; however, it is recommended that the lower wattage heaters be reinstalled when using the more common fluids that boil between 413 and 420°F. Also, to avoid excessive heater temperature, the fluid level should be a minimum of 1/2 in. above the heaters and certainly never allow exposed heaters since this will eventually cause heater failure or reduced heater life.

Normally, batch systems incorporating the use of electrical immersion heaters are equipped with two types of cycles—idle and reflow. This is necessary to avoid operating the system continuously at full power, which can cause additional vapor loss and fluid decomposition.

A typical soldering cycle is shown in Fig. 10.12. Here we have four 30-s reflow cycles and five idle cycles. To simplify how this works, once work or assemblies are loaded on the elevator platform, the reflow switch is activated, which automatically increases the heater power. Once the reflow cycle is completed, the system automatically reduces the heater power input to about 25 percent of the full power, thus conserving energy and operational costs.

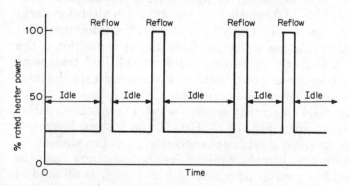

Figure 10.12 Idle and reflow cycles as they relate to heater power operation.

10.2.3.4 Vapor condensing coils. In the conventional batch system, the condensing coils are located in a helical configuration along the vertical sides of the tank. Normally, there are just two sets of coils, called the primary and secondary coil. Their location determines the size of the secondary and primary vapor zones, which can vary depending on the size of the tank or vapor phase system used.

The primary coil is the one nearest the boiling fluid in the bottom of the tank. The function of the primary coil is to condense only the primary vapor. In order to do this, the coolant within the coil (typically heated water) must flow at a temperature lower than the primary vapor temperature but higher than the secondary vapor temperature. Since the difference in the two vapor temperatures is large, it is simple to accomplish this by allowing the water supply or inlet to range between 100 and 120°F and the outlet to the primary coil range between 160 and 180°F. These temperature readings of the water inlet and outlet for the primary coils will provide acceptable operating conditions which are producing sharp vapor and temperature gradients between the two vapor zones.

The secondary coil should operate such that the coolant is supplied at 45 to 65°F. In any case, preferably the coils should not run at a temperature below the dew point because this will cause water condensation and promotion of additional acid formation. As mentioned, the secondary vapor zone is critical to the economics of this system since it is used to reduce losses of the expensive primary fluid. It does this by reducing the sharp temperature gradient between the primary vapor and air, which prevents aerosol formation above the interface. Furthermore, it also provides a flow pattern to assist in carrying the primary vapor to the coils.

10.2.3.5 Fluid treatment system. Regardless of the fluids used in a vapor phase system, it is necessary to have a fluid treatment system which neutralizes any acid formation. As mentioned, hydrofluoric acid can be formed from the low-level decomposition of the fluoroinert fluids, and hydrochloric acid can be formed from the decomposition of the secondary fluid. A properly operating and designed fluid treatment system is necessary to prevent acid buildup, which can attack the metals in the system and even on the assembly.

A well-designed fluid treatment system, which is sometimes called a *scrubber,* is illustrated in Fig. 10.13. Here we have three basic sections, the first (1) of which is where the secondary fluid condensate is drained into a funnel within the tank under the secondary coils. In this same section, fresh water at a rate of 1 to 2 gal/h is allowed to enter from the bottom of the section. Since water is lighter than the secondary fluid and nonmiscible, it forms water droplets which rise

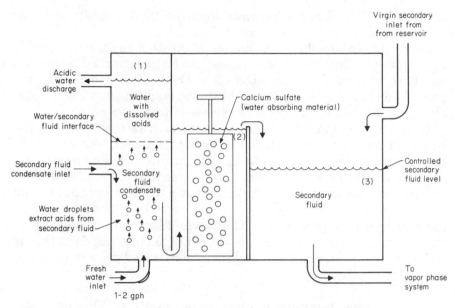

Figure 10.13 Typical fluid treatment systems for conventional vapor phase reflow systems.

through the primary fluid condensate and floats to the top of the secondary fluid level. To maintain a continuous flow of water or continuous acid neutralization, the water which extracts the acid from the condensing fluid is continuously drained out of the system.

The second section (2) now contains neutralized secondary fluid which could have some absorbed water. To remove any absorbed water, a drying column of calcium sulfate or molecular sieves is used. Whichever type of drying column is used, the columns must be replaced with fresh ones in order to avoid columns saturated with water, and thus unable to remove absorbed water from the secondary fluid.

The third section (3) is where the treated secondary fluid is collected from the overflow from section (2); section (3) acts as a controlled reservoir so that there is always a continuous flow of secondary fluid into the system.

10.2.3.6 Flux filtration system. The most commonly used fluxes for vapor phase soldering are rosin-based. During the reflow process, significant amounts of rosin on the assembly are dissolved and washed off into the boiling fluid. As more assemblies are soldered, the amount of dissolved flux in the boiling fluid increases to a point approaching saturation, depending on the type of fluoroinert fluid used. In any case, when the system is shut down and the fluid is allowed to cool to room temperature, the flux can separate from the fluid, solidify, and condense on the electrical immersion heaters.

Once flux is allowed to condense on the heaters and the heaters are turned on, the flux quickly decomposes and acts as a thermal insulator. This, in turn, can reduce the life of the heaters and produce excessive thermal decomposition of the fluids, resulting in a loss of valuable fluid as well as an increased release of toxic vapors.

To avoid these concerns, a pump is used to draw the fluid from the bottom of the tank and through a filter, returning the filtered fluid to the sump. For preferred conditions, a 1-μm filter should be used and replaced regularly depending on usage. In any case, the filtered fluid must not be pumped through the system at a temperature that will be harmful to the pump. In some instances, continuous filtration during machine operation is necessary. This type of filtration is more complex than those that are used after machine usage since the fluid that is pumped must be cooled down and then be reheated prior to return to the sump.

10.2.3.7 Thermal mass batch vapor phase systems. Thermal mass batch systems are designed and function differently from the conventional batch systems. The two major differences are the vapor condensation and generation techniques.

In this system, as shown in Fig. 10.14, the primary vapor is not condensed with a water-cooled condensing coil as in a conventional sys-

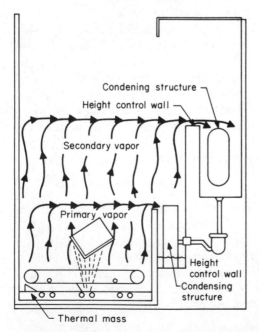

Figure 10.14 Thermal mass batch phase system. (*Courtesy Corpane Industries.*)

tem but with a multivapor separation column–condenser eliminating
the use and possible discharge of water. Furthermore, the secondary
vapor is condensed using a cooling coil that is temperature-controlled
with a closed-loop, efficient refrigeration system. Upon further review
of this type of vapor-condensing scheme, the manufacturer proposes
that the sharp vapor and temperature gradients necessary for minimal vapor losses are enhanced by locating the condensing structures
behind two walls in the tank. This design has been shown to produce
effective gradients by drawing the vapors over the side by a partial
pressure differential in order to maintain vapor height control and
condensation.

As mentioned, the other unique feature is the technique by which
the vapors are generated. Instead of using electrical immersion heaters, a molten bath of tin-bismuth is used as a heat source to produce
the vapors. To accomplish this, droplets of fluids are allowed to land
on the molten metal surface, which produces rapid vapor generation
and maintains a saturated vapor zone. Producing vapor in this manner offers several advantages. One of these is simply not having to
purchase the large volume of fluid that is necessary for conventional
systems, thereby significantly reducing the costs of machine start-up.
Another advantage is that there is no need to filter the fluids caused
by the runoff of flux into the sump. In this case, when flux does run off
the assembly into the sump, it is skimmed away from the top of the
molten alloy and, of course, never dissolves in the boiling fluid since
there is none. Last, having no electrical immersion heaters removes
the concerns of the effects decomposed flux residues have on the heaters and the additional low-level decomposition of the fluids due to continuous boiling.

10.2.3.8 Conveyorized vapor phase systems. Conveyorized vapor
phase systems are typically used for manufacturing high-volume
SMAs and even fusing multilayer PCBs in the fabrication industry.
Hybrid Technology Corporation (HTC), which developed the first
conveyorized system in the late 1970s, is now one of several manufacturers of conveyorized units. Even though these systems can have
fewer vapor losses than batch systems, it has been proved that there
can be significant differences between one manufacturer's design and
another's as it relates to vapor losses in conveyorized systems. The
buyer should also beware since some systems tested have shown excessive fluid dragout and are thus uneconomical for high-volume usage.

As with batch systems, before purchasing a vapor phase system, it
is important to become familiarized with the system's design since
small design differences can have a significant effect on the machine's
effectiveness and efficiency. Although all available system designs

cannot be discussed, the basic principles of designs can be reviewed as shown in Fig. 10.15.

Figure 10.15 illustrates a single vapor, conveyorized system. Most conveyorized systems use a lightweight belt with low thermal conductivity which carries the assemblies down an inlet throat and through a vapor zone that is approximately 24 in. long. The rate of the belt determines the length of the reflow duration at the elevated temperature.

The vapor zone is created with electrical immersion heaters and is condensed with water-cooling condensation coils. Due to the tank closure at the top, which is unlike vertically loading batch systems, and condensation coils on both inlet and exit throats, a secondary vapor zone acting as a vapor blanket is not a mandatory condition. However, conveyorized thermal mass systems which condense and generate vapors in the same manner as thermal mass batch systems employ a

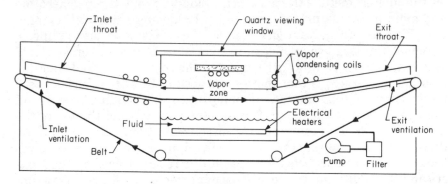

(a)

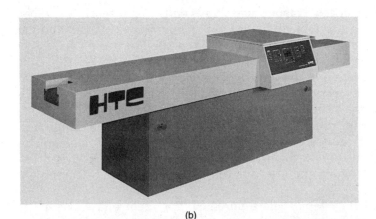

(b)

Figure 10.15 Conveyorized vapor phase reflow system (conventional single vapor type). (*a*) Cross-sectional view; (*b*) external view. (*Courtesy Dynapert-HTC.*)

dual vapor zone in the conveyorized system. Usage of the conveyorized thermal mass (dual vapor) system shows that even lower primary fluid losses are obtainable; however, losses of the secondary vapor may nullify the losses saved with the primary fluid unless modifications are incorporated to reduce the high secondary vapor losses.

10.2.4 Vapor phase reflow soldering concerns

The vapor phase reflow process can be considered one of the more complex reflow techniques, at least on a par in complexity with wave reflow soldering. From a manufacturing standpoint, several areas must be continuously controlled and understood in order for this process to be efficient and economical. These are as follows:

1. *Fluid type:* The fluoroinert fluids used in this process cost between 550 and 610 U.S. dollars per gallon, and therefore large losses of this fluid make a very uneconomical process. It is conceivable that a poorly maintained or operated system could lose as much as 1 to 2 gal of fluid per 8-h operation period.

   The first concern in minimizing fluid losses is to use a stable fluid or one that has very low level amounts of decomposition. Several of these fluids exist, as discussed in Sec. 10.2; but in all cases, the most stable fluid is composed of only carbon and fluorine atoms arranged in a primarily cyclic configuration. Other fluids, which may contain elements such as nitrogen and oxygen, have been shown to be significantly less stable but slightly less in cost.

2. *Machine type:* The type and manufacturer of the vapor phase reflow system is also a major factor regarding the amounts of fluid lost. Although all machine manufacturers claim that their systems are economical, in reality there are some cases where this does not hold true. For example, based on actual testing, there are some conveyorized systems that when used in low- to moderate-volume production, are economically efficient. However, when the same machine is used in volume production, large quantities of fluid dragout make this system very uneconomical. Because the effects of fluid losses differ among the various system designs and manufacturers, it is recommended to test the system prior to purchase in the same way it is going to be used in production.

   Under well-controlled and well-maintained conditions, the expected operating costs for fluid losses are as shown in Fig. 10.16. Analysis of these data shows that conveyorized systems are more economical than batch systems. This is due, in most cases, to the smaller square area of the opening through which the work travels.

3. *Process parameters:* The process parameters can affect the time it takes the solder to reach its liquid state, how well the solder met-

Figure 10.16   Operating Costs due to Fluid Losses for Vapor Phase Reflow Systems

	HTC (Dynapert) model designation						
	2424*	1826*	1416*	1214*	912*	IL-24†	IL-16†
Operating costs:							
Aperture area, ft^2	4.92	3.83	2.17	1.70	1.40		
FC-70 (0.05 lb/h ft^2 at \$38.10/lb), \$/h	9.37	7.30	4.13	2.67	2.04	5.32	2.78
FC-113 (0.5 lb/h ft^2 at \$1.08/lb), \$/h	2.66	2.07	1.17	0.76	0.58		
Projected solvent loss, \$/h	12.03	9.37	5.30	3.43	2.62		

*Batch systems.
†Conveyorized systems.

allurgically reacts with the terminations to be soldered, and the amount of fluid losses to the environment. Upon setting up the vapor phase reflow process, one should consider the following:

  a. The fluid level should be above the electrical immersion heater and should be filtered clean. This will prolong heater life and promote uniform boiling of the fluid.

  b. The cooling coils should always be operating at the manufacturer's recommended temperatures. It is recommended that thermocouples be installed to be able to continuously monitor the inlet and outlet temperatures of the coils.

  c. The vapor temperatures should be continuously monitored along with the boiling point of the fluid.

  d. The dwell time of the work in the primary zone is dependent on the mass of the assembly. The greater the mass, the longer the dwell time in the primary vapor. As a starting point, the typical assembly that is moderately populated may see a 30-s dwell time. The same assembly with a laminated thickness greater than 0.062 in., having inner metal planes greater than 1 oz, component heat sinks, or a thick metal core, may require a 45- to 90-s dwell time. The general rule in deciding the dwell time is to use the minimum time possible to achieve uniform metallurgical results for all components to be soldered.

4. *Acid neutralization:* No matter what system or fluid is used, there must be continuous fluid treatment in order to remove acids that form during the low-level decomposition of the fluid. Although most systems incorporate an acid neutralization system, close monitoring is necessary.

5. *Machine maintenance:* Without exception, all systems require routine and continuous maintenance. This should consist of the following:
   a. The electrical immersion heaters should be removed and scrubbed clean of any carbonous deposits such as burnt rosin flux.
   b. Depending on frequency of use, the fluid should be pumped through a 1-μm filter, neutralized with deionized water via separatory funnel techniques, and dried with a desiccant. When the fluid is very contaminated, the contaminants can be almost completely removed via distillation techniques.
   c. Wipe the cooling coils clean. This will allow more uniform and complete vapor condensation and reduced vapor losses.

10.3   Infrared Reflow Soldering

It is quite clear throughout the industry that soldering techniques such as vapor phase, infrared, and dual wave differ. However, it is not clear, although it actually exists, that there are various infrared reflow techniques which transfer heat differently. This difference has to do with the wavelength of radiation exposed to the work to be reflowed.

Infrared radiation is a particular band of electromagnetic radiation that falls between light and radio waves. This is illustrated in Fig. 10.17. It is the same type of radiation as light, but has a longer wave-

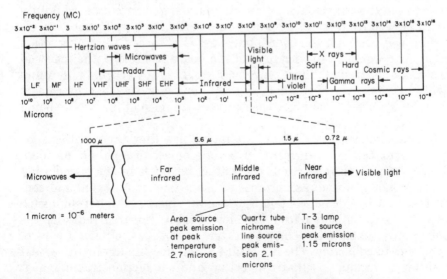

Figure 10.17   The location of the infrared region in the electromagnetic spectrum.

length or lower frequency. This radiation should be thought of as being made up of energy packets or photons. The photons of a given wavelength or frequency all have the same energy in each photon. The number of photons striking a surface in a period of time determines the intensity of radiation incident on that surface.

Any object emits radiation, in that, as the temperature of an object changes, the type and amount of radiation change. An ice cube emits radiation most strongly at 10.6-μm wavelength; the intensity of radiation is low, and the wavelengths are very long and far away from the visible. Because the peak wavelength of the radiation is not dependent on the material that is emitting it, the metal tray that holds the ice cube is also emitting radiation most strongly at 10.6 μm. If the tray were heated to 600°C, it would begin to glow dark red. As objects are heated, the wavelengths they emit become shorter and emit enough visible light (red) for the eye to detect. This sample relationships between temperature, wavelength, and total radiation hold for all objects as they are heated.

When an object emits radiation, it travels through space until it encounters something that can absorb it. When a photon is absorbed, it surrenders all of its energy to the absorbing material. Some materials absorb a given wavelength strongly, while others may reflect that same wavelength. The absorptivity of a material is the measure of how much radiation will be absorbed as it passes through a given thickness of the substance. If made thick enough, any material will become opaque. The behavior of material with visible light gives good clues as to how that material will absorb infrared radiation. It is possible to see the traces on one side of a glass-epoxy circuit board faintly when viewed from the other side. Some of the light (radiation) travels all the way through the board, but most is absorbed. This absorption occurs throughout the thickness of the board; therefore, the material is semitransparent at these wavelengths. A black-anodized aluminum heat sink, on the other hand, is very opaque even when extremely thin. This material is highly absorptive, with most of the energy being absorbed on the surface.

Convection, conduction, and radiation are the three methods by which heat can be transferred from one object to another. As mentioned, convection requires direct contact between an object and a hot vapor or hot gas such as in vapor phase reflow or an oven, and conduction requires direct contact between an object and a hot solid such as when using a solder iron to reflow a joint. Infrared radiation requires only that the hot object (emitter) be within the field of view of the object to be heated. The important difference among the three heat transfer techniques is that convection and conduction transfer heat from the surface of the object to be heated, whereas true infrared ra-

Figure 10.18 Properties of Various Materials at Peak Emission Wavelengths Emitted by Area and Lamp Infrared Emitters

Material	Area panel emitters (wavelength, 2.7 μm)	Lamp (T-3) emitters (wavelength, 1.15 μm)
Glass-epoxy (FR-4, G-10, F-12)	Absorptive	Semitransparent
Polyimide-glass	Absorptive	Semitransparent
Solder flux, solder paste	Absorptive	Semitransparent
Anodized or oxidized metal (Al, Cu, Ni, Sn, Pb)	Absorptive	Absorptive
Polished metal (Al, Cu, Ni, Sn, Pb, Au)	Reflective	Reflective
Ceramic (Al$_2$O$_3$, SiO$_2$, BaTiO$_3$)	Semitransparent to absorptive	Transparent to semitransparent
Black plastic component packages	Absorptive	Absorptive

diation delivers heat directly to the interior of the object. Rapid heating by infrared energy at wavelengths where the object is transparent or semitransparent allows the interior and surface of the object to be heated at nearly the same rate.

In order for infrared energy to penetrate and deliver heat to the interior of an object and avoid excessive surface heating, the material must be transparent or semitransparent to the wavelength of the heat source. The two types of heat sources typically available in infrared reflow equipment consist of either an area or lamp source emitters which can generate radiation at wavelengths of 2.5 to 5 and 1 to 2.5 μm, respectively. Figure 10.18 compares the properties of materials at the range of wavelengths emitted by area and lamp emitters. Figure 10.18 shows that area source emitters generate radiation which is mostly absorptive and, therefore, produce mostly surface heating in contrast to the lamp emitters, which deliver heat directly to the interior of most materials.

From this analysis it can be deduced that infrared reflow systems utilizing area source emitters generate heat to the product much like that of an oven, which is a convection-dominating system. Systems with lamp emitters are truly infrared reflow and, therefore, generate heat which is infrared-dominating.

10.3.1 Area source panel emitters

Area source panel emitters are constructed to operate on the secondary emission principle and are the most common type of emitter used

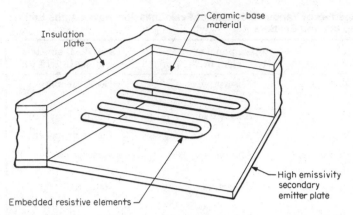

Figure 10.19   Area source panel emitter with resistive elements embedded in a ceramic base material.

for soldering applications. As shown in Fig. 10.19, this is accomplished by embedding resistive elements in appropriate, thermally conductive ceramic-based materials in close proximity to the intended emitting side of the panel. A backing of high R value insulation is affixed behind the ceramic element mass to ensure emission in one direction only. A thin, low-mass, electrically insulative, high-emissivity emitter material is then attached to the front (emitting) side of the panel. In operation, the resistive element heats the flat secondary emitter material, which in turn emits diffused radiation uniformly across its entire surface.

Another type of area emitter, as shown in Fig. 10.20, is constructed so that the resistive element is embedded between a reflection and radiation plate instead of within the ceramic-base material. This type of construction usually reduces the cost of the emitter; however, it is possible for the radiation plate to be heated nonuniformly and therefore emit radiation nonuniformly.

The area emitters normally have a peak temperature rating of 800°C and a typical operation lifetime of 4000 to 8000 h. Cleaning of

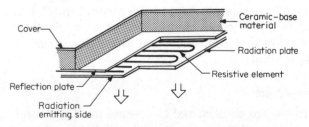

Figure 10.20   Area source panel emitter with resistive element embedded between a reflection and radiation plate.

these emitters may be necessary due to the condensation and decomposition of flux volatiles on the emitting side of the panels.

10.3.2 Lamp source emitters

The two commonly used lamp source emitters are the T-3 lamp and the Ni-Cr quartz tube lamp. The T-3 lamp consists of a sealed tubular envelope enclosing a helically wound tungsten filament supported by small tungsten disks. The tube is evacuated and filled with an inert gas such as argon to reduce oxidative degeneration of the filament and sealed. This lamp has a peak temperature of 2246°C.

The Ni-Cr quartz lamp is similar to the T-3 in construction, except that the filament is contained in a nonevacuated quartz tube. This lamp has a peak temperature of 1100°C.

The quartz-tungsten lamps are rated for 6000 h of life at chamber temperatures of 1000°C. The life is limited by the tendency of vitreous quartz to crystallize at these temperatures. At solder reflow temperatures, the life of these lamps is enormous, with a predicted lamp life in excess of 1 million h at the applied voltages used in the reflow soldering process.

10.3.3 Infrared reflow soldering systems

Figure 10.21 illustrates a typical infrared reflow system. The process area is constructed using the area emitter panels or lamps and various types of refractory insulation as the internal tunnel lining medium. No matter who the manufacturer, a conveyor traveling through the tunnel consists of either the mesh or meshless type. The meshless type holds only the edges of the board being processed with no contact with the bottom side of the PCB.

The first segment of the system, from the entry area, incorporates a short preheat zone which is designed to drive off flux volatiles in the solder paste to minimize solder splashes and part movement. The next segment may consist of a vented transition area where these volatiles are removed. This area connects directly to the process zone, which typically consists of six (min.) individually controlled zones of top and bottom emitters.

Furthermore, inert gas can be introduced into the process area through stainless steel tubes embedded in the tunnel insulation. Inert gas is sometimes used to reduce the amount of oxidation which occurs with the solder paste, terminations, and component leads.

10.3.4 Infrared reflow process concerns

The infrared reflow process is the simplest to operate and maintain when compared to vapor phase and wave reflow soldering techniques. The only two major process parameters are the emitter temperatures

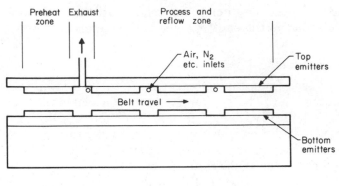

(a)

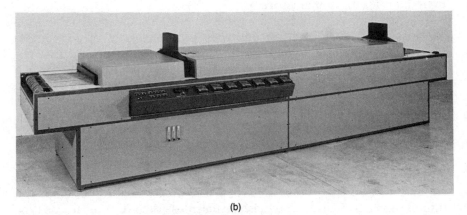

(b)

Figure 10.21 Diagram of a conveyorized infrared reflow system. (a) Cross-sectional view; (b) external view. (*Courtesy Vitronics, Inc.*)

and belt speed. With these two variables one can change the temperature profile of the assembly as it passes through the system to a point where a numerous number of profiles can be established.

One might then ask which temperature profile is the preferred one? With this in mind, there are certain conditions which must be considered as follows:

1. The initial preheat zone should not expose excessive radiation to the solder paste but a sufficient amount of radiation to remove the volatiles from within the solder paste.

2. The preheat zone should be long enough to minimize nonuniform temperature at the substrate surface.

3. The elevated preheat zone prior to the reflow zone should not overheat the substrate so as to cause damage such as resin charring and measeling.

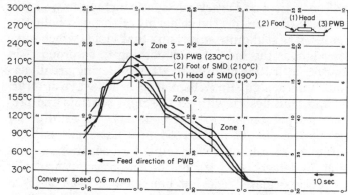

Figure 10.22 A typical temperature profile of an assembly reflowed via infrared. Notice the three distinct temperature zones.

4. The elevated preheat zone should bring the areas to be soldered up to the wetting temperature of the solder alloy.

5. The reflow zone should be spiked so that the assembly is exposed for a minimum amount of time at the spiked temperature but for a sufficient length of time to produce a uniform reflow of solder.

6. The conveyor should be set at a speed which satisfies the manufacturing throughput requirements. If faster conveyor speeds are used without satisfactory soldering results, it will probably be necessary to increase the length of the reflow system with the addition of radiation emitters.

With the above conditions taken into consideration, let us review a typical temperature profile for infrared reflow as shown in Fig. 10.22. The best results will be achieved when the system is set up with three different and distinct temperature zones. Zone 1 establishes elevated temperature for all the materials composed in the assembly and also initiates the evaporation of solvents from within the solder paste. Zone 2 is utilized to activate the flux within the solder paste to achieve solderability upon reflow and to bring the entire assembly to higher preheat temperatures to prevent thermal shock during exposure to reflow temperatures. Zone 3 provides sufficient radiation to bring all the metals to be joined to wetting temperatures and to provide a rapid spike in temperature to bring the solder paste to its molten phase.

10.4 Dual-Wave Reflow Soldering

The wave soldering technique which utilizes pumped waves of molten solder is a well-established mass soldering process in the PCB industry. Although the theoretical and practical aspects have been well documented over the past 15 years for the soldering of through-hole conventional assemblies, it is still quite surprising that yields which have been observed throughout the industry are nothing to be proud of in most cases. When the multitude of areas in wave soldering are understood, such as machine design, flux chemistry, board-wave interaction, metallurgy, and PCB design with all the right controls and process techniques incorporated, the defect levels, especially for conventional assemblies, should not exceed fifty parts per million. Simply stated, for every 1 million solder joints formed, there should not be in excess of 50 defects. And when a defect occurs, it typically should not be a process-related one but one such as lead or PCB solderability, which is usually the major problem when process problems are eliminated.

Unlike in conventional through-hole assemblies, SMAs can have a substantially more complex variety of leaded and leadless components. Although the same technical principles apply when wave soldering the variety of surface mount components, the interactions between the molten wave of solder, the components, and the PCB are different and can be considered more complex than through-hole technology. Board entry, heat transfer, capillary forces, hydraulic forces, lead-to-hole clearance, and assembly exiting conditions are some of the major concerns for this standard construction, as shown in Fig. 10.23. The only typical construction variables with through-hole assemblies are lead-to-hole clearance and lead protrusion length, both of which are normally understood by the ways in which they have an effect on yield.

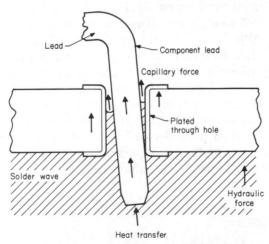

Figure 10.23 The interaction between a solder wave and plated through-hole-mounted component.

The interactions between the molten solder wave and the more complex variety of leaded and leadless surface mount component terminations are the major differences which impose mandatory changes relating to molten solder wave formation, PCB design, component layout, and process parameters. There are five major types of surface mount terminations that must be contended with in the wave soldering process: the dual leadless chip components (capacitors, resisters, etc.), multileadless chip carriers, J-leaded SOICs, gull-leaded SOICs, J-leaded PLCCs, and the gull-leaded flat packs. Five major variations in component terminations can therefore occur on the same assembly that is to be wave soldered. Certainly this makes wave soldering of all varieties of surface mount components magnitudes far more complex than plated through-hole technology.

Surface mount dual leadless components such as chip resistors and capacitors are by far the easiest to solder, and zero defect yields can easily be reached when all process and design factors are considered. However, when soldering active surface mount components, especially when leads are on all four sides of the package, the complexities involved in adding and removing the correct volume of solder from each lead are difficult, and in some cases that are more dependent on machine design, next to impossible. Certainly, unlike in wave soldering chips and possibly SOIC packages, the development of soldering multileaded packages with leads on all four sides can still be considered at its infancy stage. The contributing factor which typically increases the difficulty of soldering the more complex surface mount components without defects is the ability of removing the excess solder between adjacent leads, which normally end up as solder bridges. The areas where solder bridges occur are usually the trailing leads that exit the solder wave. New soldering systems which incorporate special wave forming features and techniques have all been tried by the author. Machine manufacturers who sometimes claim their system's ability to produce high yields when soldering multileaded surface mount components have been overestimating their machine's capabilities.

10.4.1 Dual-wave reflow soldering systems

Wave soldering systems are available with various levels of sophistication and capability, ranging from small laboratory machines to computer-controlled high production systems. In any case, the heart of the wave soldering system is the solder bath and the manner in which the molten bath of solder is formed. When soldering SMAs, the typical bath design consists of two independent molten baths of solder which are commonly identified as the turbulent wave and smooth wave. The

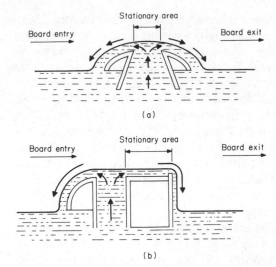

Figure 10.24 Symmetrical and asymmetrical waves commonly used in standard conventional and dual-wave soldering systems.

systems that employ these two types of waves are known universally as dual-wave soldering systems.

In the early stages of wave soldering SMAs which consisted of chip resistors, chip capacitors, and SOT 23s, it was found that the conventional wave soldering system designed with either an asymmetrical or symmetrical wave could effectively solder these types of assemblies. These two types of classic waves are illustrated in Fig. 10.24.

The asymmetrical wave consists of a chimney through which molten solder is pumped upward. As it reaches the top, it flows in two areas or directions. The area which the board enters first is the fastest moving portion of the molten solder. The other area from which the board exits is an almost stationary area of various lengths depending on the manufacturer's specific chimney design.

The symmetrical wave, which is the less popular of the two, consists of molten solder pumped upward through a chimney in the same manner as in the symmetrical wave. The variation in this chimney design causes the molten solder to flow rapidly toward both the entry and exit areas of the board from the wave, with the center of the wave being the most stationary area.

Further analysis of wave soldering assemblies with chips and SOTs utilizing either one of the two conventional waves revealed that certain restrictions occurred. The major restriction under these soldering conditions is the layout or orientation of the components. When lands are designed correctly, as discussed in Chap. 6, high-yield soldering can be obtained when both chips and SOTs are oriented on one axis parallel to the wave, as shown in Fig. 10.25a. Figure 10.26 illustrates

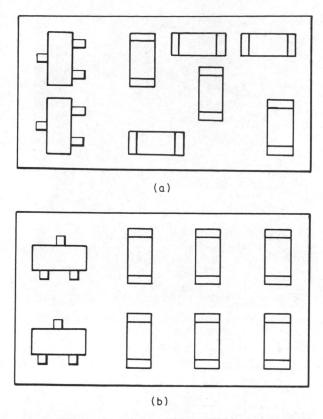

(a)

(b)

Figure 10.25 Chip and SOT orientations on a PCB. (*a*) One axis of orientation parallel to the wave movement; (*b*) several axes of orientation.

an actual high-volume assembly with chip capacitors soldered with only an asymmetrical single wave system.

When the components are on several axes of orientation, as shown in Fig. 10.25*b*, wave soldering chips and SOTs utilizing a single wave become ineffective due to the solderless "shadow" effect. This solderless shadow is simply the absence of solder application on a component's termination due to diversion of solder around the termination. When such assemblies are processed through a single wave, the components themselves, when making contact with the molten wave, produce a wake of solder on the trailing edges. The area just behind the trailing edges is where the solder "shadow" occurs and is illustrated in Fig. 10.27. In most cases, it can be assumed that components cast a 45° solderless "shadow" at their trailing edges. This means that exceptionally tall components such as tantalum capacitors require their solder terminations, particularly at the trailing edges, to extend

Figure 10.26 High-volume assembly soldered with only an asymmetrical wave. All chip capacitors are oriented parallel to the wave.

Movement of board into wave

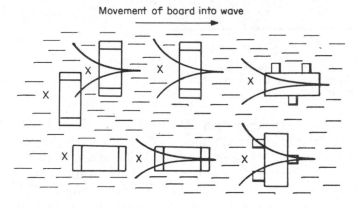

Figure 10.27 Solder shadowing occurring in a single "smooth" wave.

far enough beyond the shadow zone to capture solder from the passing wave.

In order to overcome these design restrictions as they relate to component orientation and density, the dual-wave soldering system was developed. In most cases, the dual-wave system consists of two sections. The first section, the turbulent wave, overcomes the possible complex component geometrics and orientations by scrubbing the surface of the assembly with the molten solder in all directions and thereby overcoming solder shadowing effects. Upon exiting the turbulent wave, the terminations will have excess amounts of solder. This is caused primarily by the rapid velocity of the wave and the angle at which the turbulent solder is drawn off the board as it exits the wave. To rectify these concerns, the assembly enters the second wave, which effectively removes the excess solder, leaving the assembly with all joints wetted and free from shorts or bridges.

There are three major types of wave systems that are used to solder SMAs. Each type of wave system, as illustrated in Fig. 10.28, has a different degree of effectiveness when soldering assemblies with various densities of chips and SOTs. To date, none of these dual-wave systems has proved very effective in soldering surface mount ICs, especially packages with leads on all four sides.

10.4.1.1 Narrow symmetrical turbulent wave. This dual-wave soldering system, as shown in Fig. 10.28a, consists of a very narrow, slotted chimney which produces a rapidly moving, turbulent, symmetrical wave and an asymmetrical wave. This turbulent wave section can overcome the formation of solder "shadows," as discussed earlier, provided the component density is low to moderate. When component densities are high, the turbulent section may require modification to overcome shadowing and solder bridging. The major cause for these concerns with higher component densities is that the turbulency of the molten solder is "irregular" along the narrow, slotted chimney, in that the amount of turbulency can vary along the slot even at a consistent pumping speed.

10.4.1.2 Perforated oscillating turbulent wave. This dual-wave soldering system creates a turbulent wave unlike that of the narrow symmetrical turbulent wave, although the second wave in this system produces an asymmetrical waveform. Solder turbulency is created by pumping molten solder through a perforated nozzle which can be adjusted so that it is stationary or moving at various adjustable speeds in a direction that is parallel to the second wave. The nozzle consists of a hollow metal tube with staggered rows of holes. The holes, which are available in various diameters, are the exiting areas for the molten solder. Upon exiting these holes, the molten solder "bubbles" in a consistent and regular pattern, depending on the speed of the nozzle and the rate at which molten solder is pumped through the holes. Figure 10.28b illustrates this type of system design.

This type of dual-wave design offers the greatest flexibility when soldering lowly to highly populated assemblies. The major advantage with this dual-wave design is the ability to scrub the assembly at different rates. This added capability can usually overcome the occurrence of solder shadowing, which can be caused by high-profile parts and high-density assemblies. Figure 10.29 illustrates an actual, perforated, oscillating turbulent wave while operating and not operating.

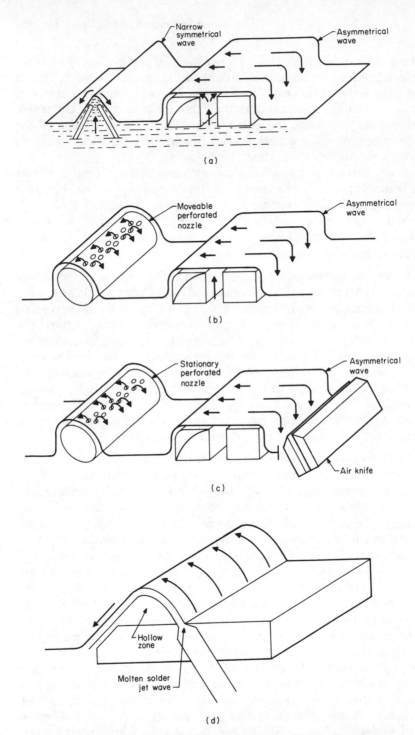

Figure 10.28 The four most common types of dual-wave designs. (*a*) Narrow symmetrical; (*b*) perforated oscillating; (*c*) perforated stationary; (*d*) jet wave.

(a)

(b)

Figure 10.29 A perforated, oscillating, turbulent-wave system. (a) Operating; (b) not operating. (*Courtesy Sensby Corporation.*)

10.4.1.3 Perforated stationary turbulent wave. The stationary perforated turbulent wave operates under the same principles as the "movable" type, except that the perforated nozzle is in a fixed location. This system (see Fig. 10.28c) also has an asymmetrical wave with a hot air knife just beyond where a board would exit the second wave. The hot air knife is an optional item with this type of system; however, it does deserve some discussion.

The hot air knife was primarily designed to remove solder bridging between through-hole-mounted component leads on the solder side of the board after just exiting the wave. The major causes of solder bridging are PCB design, process control, and not understanding the chemistry of wave soldering. To overcome these deficiencies, companies have adopted the air knife, which utilizes high-velocity heated air to blow away the solder bridges. The concerns regarding this technique as it relates to board reliability are beyond the scope of this discussion; however, using the hot air knife in the hope of effectively soldering surface mount ICs has been examined.

The major problem when soldering leaded surface mount components is solder bridging between the trailing leads that exit the wave. Upon analysis, when used to remove these bridges, the air knife was ineffective. Higher velocity air was required to account for the added distance between the air nozzle and board surface caused by component profiles. The higher air velocities were found to remove too much solder and at the same time were unable to consistently remove all solder bridging between leads. Furthermore, the risk of blowing off chip components became an increasing factor due to the weakened adhesive bonds after reflow and additional stresses imposed on the bonding adhesive from the high-velocity, high-temperature air.

10.4.1.4 Jet wave. The jet wave is unlike any other waveform. In fact, it is neither a dual-wave nor turbulent-wave system. It employs a high-velocity molten solder wave which flows in only one direction, as shown in Fig. 10.28d. Due to the wave's velocity and shape, a hollow zone is created under the molten solder wave. This technique substantially increases the hydraulic forces that are exerted against the passing assembly along with some turbulency created when components enter and exit the hollow zone. This technique is capable of soldering assemblies with chips and SOTs; however, like all the other types, it is ineffective in soldering leaded surface mount components.

10.4.2 Dual-wave reflow soldering process concerns

After operating a dual-wave soldering system, it is immediately obvious that there are three areas that greatly affect the soldering yields. These consist of the following:

1. PCB design
2. Flux type and usage
3. Preheat temperature profile

Any one of these areas that is not incorporated and maintained correctly can cause massive solder defects, which usually consist of solder opens and shorts. Furthermore, losing SMDs that are bonded to the substrate can be another type of defect produced from an improperly set up wave soldering process. In the continuing sections, let us review these areas as they relate to soldering yields.

10.4.2.1 The PCB design as it relates to dual-wave soldering yields. Unlike in vapor phase and infrared reflow soldering, the board design can have the greatest effect on obtaining high or low soldering yields. *No matter how well the soldering process and materials are set up, in most cases they will not overcome the defects caused by board design inefficiencies.* Proven board designs are provided in Chaps. 5 and 6; however, let us produce a basic list of rules that should be followed specifically for assemblies that are dual-wave-soldered. The basic board design rules for dual-wave soldering are as follows:

1. Whenever possible, use a minimum substrate thickness of 0.062 in. Thinner substrates tend to sag during their exposure to the second wave and can literally dive into the molten bath of solder. Fixtures may correct this problem; however, they tend to be very troublesome in high-volume production.

2. Use a temperature-resistant solder mask, especially when large ground planes or wide conductors are on the solder side of the board. These solder masks are neither liquid nor dry film but liquid photoimagable, as discussed in Chap. 5. Due to their higher temperature resistance and adhesion at soldering temperatures, the liquid photoimagable solder masks will not wrinkle over tin-lead-plated conductors and will provide superior adhesion when exposed to several dual-wave cycles.

3. Design the solder mask artwork so that when processed, the solder mask will not cover the lands. If liquid solder mask is used, the artwork should be enlarged by a minimum of 0.010 in. If dry film or liquid photoimagable solder masks are used, the artwork should be enlarged by a minimum of 0.005 in.

4. Do not locate quad packages with J or gull-wing leads on the solder side of the board when the assembly requires wave soldering. Currently, there is no process that can prevent the formation of solder bridges between adjacent leads on these types of components. The

hot air knife has been conclusively proven ineffective in removing solder bridges located on these types of components.

5. Locating SOICs on the solder side of the substrate is acceptable. This is not a preferred design rule, but high yields can be obtained if the proper land designs are used as given in Chap. 6.

6. Do not locate high-profile SOT 23s on the solder side of the substrate. Low-profile SOT 23s are preferred since they will allow a thinner adhesive bond layer which is less likely to flow onto the lands during curing.

7. Land geometries are one of the most critical conditions which affect wave soldering yields. Following the land geometries, as given in Chap. 5, will provide sufficient extension of the lands beyond the "shadowed" areas caused by the component as it travels through the molten wave of solder.

8. Regardless of land geometries, component layout can greatly affect the wave soldering yields. Locate high-profile components such as tantalum capacitors and SOICs so that other components are not within their solder shadow. In most cases, it is also recommended to locate tantalum capacitors on the top side of the substrate.

10.4.2.2 Solder flux as it relates to wave soldering yields. No matter how the flux is applied to the substrate—whether by foam, wave, or spray—it is the chemistry of the flux which determines its effectiveness for high soldering yields. The following are conditions that should be considered for achieving high dual-wave soldering yields:

1. When applied, the flux must uniformly coat and adhere to all surfaces of the substrate.

2. After the flux is applied to the substrate, pooling of flux is quite normal and unavoidable. Flux pooling will lead to flux droplets that must be removed. The air knife is the preferred tool for removing flux droplets. When located between areas to be soldered and allowed to reach the molten solder waves, flux droplets tend to produce solder shorts or opens.

3. The flux should become activated during the preheat stages and maintain its chemical integrity such that it does not evaporate or chemically change to a point where it can no longer reduce the surface tension of the solder once it reaches the wave.

4. The percentage of solids in a flux is critical in dual-wave soldering. Solder bridging or icicling results if the percentage of solids is too low. The major reason for this is the effect that the wave's washing action and evaporization have on the flux as it passes through the

first wave, causing an insufficient amount of flux to enter the second wave. These effects can occur more rapidly with fluxes that have 20 percent or less solids. Certain more aggressive turbulent waves may require the use of fluxes with higher solid contents.

10.4.2.3 Preheat temperature profile. Preheating the assembly activates the flux, removes extraneous volatiles from the flux, brings the metals to be soldered up to solder wetting temperatures, and elevates the temperature of the assembly in order to prevent thermal shock during exposure to the molten solder. The temperature profiles can be taken with thermal strips, thermocouples, or a pyrometer. If used and located properly, thermocouple or pyrometer measurements are preferred since a chart recorder can be attached and provide a real-time temperature profile. The actual temperature profile should sustain an increase in temperature up until 1 to 2 ft from the first molten wave where the temperature should be spiked. The surface temperatures of the substrate should vary, depending on its type and thickness and the quantity of components.

10.5 Hot Bar Reflow Soldering

The hot bar reflow soldering process is not a mass soldering technique like vapor phase, infrared, or wave soldering. Instead it utilizes metal blades which are heated by resistance or inductance. The blades, which are mounted to a head commonly called a thermode, make contact with the component leads when assembled on the board and apply sufficient temperature and pressure to reflow the solder. This technique has been predominantly used on military and aerospace assemblies with ribbon or gull-wing leaded flat packs and high I/O count, fine pitch surface mount components. The major reason for utilizing hot bar reflow for these electronic applications is to reduce the amount of hand soldering and to increase the uniformity of the solder joints. Commercial electronics have not incorporated this reflow technique, with some exceptions, since placing and soldering just one component can require as long as 10 s with a minimum time of 5 s per component for the fastest automatic systems.

10.5.1 Hot bar reflow soldering systems

Hot bar reflow soldering systems can be categorized into three types—manual, semiautomatic (see Fig. 10.30), and automatic. In all cases, the major principles of reflow are the same. First, the component is either manually or automatically picked up, aligned, and placed on

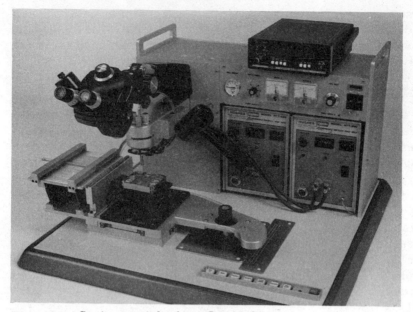

Figure 10.30 Semiautomatic hot bar reflow machine.

the fluxed land patterns with a solder thickness of 0.0003 to 0.0009 in. Second, as illustrated in Fig. 10.31, a pneumatically forced thermode with blades makes contact with the leads, which forces them against the lands if they are not already making contact. At that time a preset contact pressure, reflow temperature, reflow dwell time, and thermode

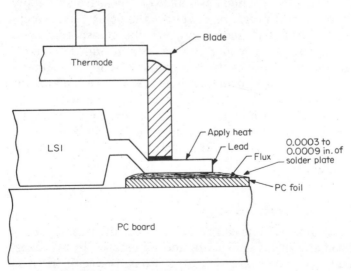

Figure 10.31 Hot bar reflow soldering technique.

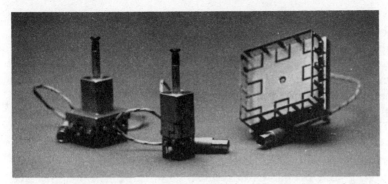

Figure 10.32   Hot bar reflow thermodes used to solder 16-, 132-, and 340-pin ICs.

lift-off temperature accomplish the soldering cycle of a component. Thermodes that hold the blades come in various sizes depending on the size of the component to be reflowed. Figure 10.32 shows three different sized thermodes which are used to solder all four sides of a 16-, 132-, or 340-lead component simultaneously. Some thermodes with one or two blades are used depending on manufacturing needs and the number of leaded sides a component may have.

10.6   Laser Reflow Soldering

Laser soldering of surface mount components has been examined and often tried without much success since the early 1970s. Problems such as charred or burned substrates, insufficient reflow, melted leads, and solder cream spattering are some of the many problems associated with laser reflow techniques. Today, however, the picture has changed. Changes in process techniques and machinery have turned what was thought to be a scientist's research tool into a potentially viable reflow and production technique.

Although laser soldering is still not a mass reflow technique like vapor phase, infrared, or wave soldering, it is now being used for reflow soldering assemblies in specific areas. One of the first areas where laser soldering has been used is with assemblies populated with leadless chip carriers mounted onto substrates containing large metal cores or heat pipes, which are commonly found in military and aerospace electronics. Assemblies such as these, which have a great deal of thermal capacity, sometimes require dwell times up to 2 or 3 min at reflow temperatures in a vapor phase system until liquification of the solder cream is completed. Necessary dwell times such as these are excessive and can certainly cause severe reliability problems with PCBs and components. On the other hand, dual-wave and infrared reflow techniques have proved ineffective in producing consistent results with these types of assemblies. Laser systems have proved effective for soldering such assemblies since the short duration pulse of the laser radiation can reflow the solder connection before the heat is conducted into the metal core or heat pipe.

Another area finding acceptance upon using laser soldering is reflow soldering the fine-pitch, high I/O count packages typically found in LSI and VLSI packaging technology. The more conventional assembly technique for these components has been using solder pastes and vapor phase or infrared reflow. Typical results include massive solder bridging and opens. Under these conditions, the bridges are typically caused from the extreme difficulty in printing uniform patterns of solder cream on lands with centerline spacings of less than 0.040 in. and the disruption and spreading of the solder pastes upon lead contact and baking out procedures. The opens between leads and land patterns can easily occur and become more prevalent as the lead count increases due to noncoplanarity of the leads. Currently no technique within this process has been developed to overcome lead noncoplanarity problems.

10.6.1 Laser reflow soldering systems

There are manual, semiautomatic, and automatic laser soldering systems. The two major concerns regarding laser soldering systems are the wavelengths of the radiation (laser scan) generated and the shape of the radiation which condenses on the termination(s). The two types of lasers commonly used for solder reflow consist of carbon dioxide (CO_2) and yttrium-aluminum-garnet (YAG) lasers. Furthermore, no matter which type of laser is used, there are typically two types of condensed radiation shapes, which can be called, for purposes of discussion, a focused beam and a diffused beam. Various machine performances can be obtained depending on the combination of conditions, as mentioned above, incorporated into a laser reflow system. Let us now review some of these concerns in the following sections.

10.6.1.1 CO_2 and YAG lasers. The mechanical designs of CO_2 and YAG are beyond the scope of this section and are not really necessary in order to describe the differences between these two. The only major difference between these lasers is the wavelength of the radiation which is released. The CO_2 laser emits radiation at 10.6 µm. As discussed in detail in Sec. 10.3 for infrared reflow, the same principles apply as to the absorbance, reflectance, and transmittance of radiation upon various materials when exposed to different wavelengths of radiations. According to tests performed by E. Lish at the Air Force

Wright Aeronautical Laboratory, the reflectance of radiation at 10.6 μm from infrared tin-lead is 74 percent, whereas reflectance at 1.06 μm is 21 percent. The practical result of these data is to show that only one-third as much energy is required from a YAG laser to melt solder as compared to the CO_2 laser. Additionally, the reflectance of printed wiring board insulation at 10.6 μm is 2 percent and at 1.06 μm is 27 percent. The practical result of these data is to show that printed wiring board material is one-fourth less likely to be burned by YAG laser energy than by the same energy from a CO_2 laser.

One conclusion that can be reached regarding the above-mentioned conditions is that the YAG laser is potentially less damaging to components and printed wiring board insulation by direct or reflected laser energy than is the CO_2 laser. Another conclusion is that the YAG laser is more efficient for solder reflow. However, these YAG laser advantages can be offset by factors such as the much higher initial costs and lower efficiency, leading to higher operating and maintenance costs than for the same power CO_2 laser.

10.6.1.2 Optic designs for laser reflow soldering systems. The way in which the optics in a laser system are constructed determines the shape of the condensing radiation on the terminations to be reflowed. As mentioned, the two most common radiation condensing shapes are the focused and the diffused beam.

The Vanzetti's laser soldering system, as shown in Fig. 10.33, has its optics constructed so that the laser radiation, generated from a 12.5-W YAG laser, is focused in a minimum spot diameter of 0.024 in. In some applications, small radiation spot sizes should be reduced to 0.004 in. in order to avoid overheating the joint and surrounding substrate material. Furthermore, this system enables the user to thermally characterize the work to be performed based on the usual conditions encountered in a solder joint for a specific application. The characterization process takes into account the geometry of the solder joint; the mass of solder, lead, and lands; the heat sinks that may be attached to the lands; and the surface condition of the termination. The system is taught to understand the typical thermal signature variation and the radiation-versus-time-profile that may be expected depending on the termination conditions. Within the capability time limits, thermal threshold schedules can be set up for each type of joint.

The NEC Corporation's optical system construction uses a different approach. Instead of focusing the laser radiation on a single spot, it diffuses the radiation so that it can condense and reflow simultaneously all the leads on two sides of a component. If the component has leads on four sides, the optical system, as shown in Fig. 10.34, is

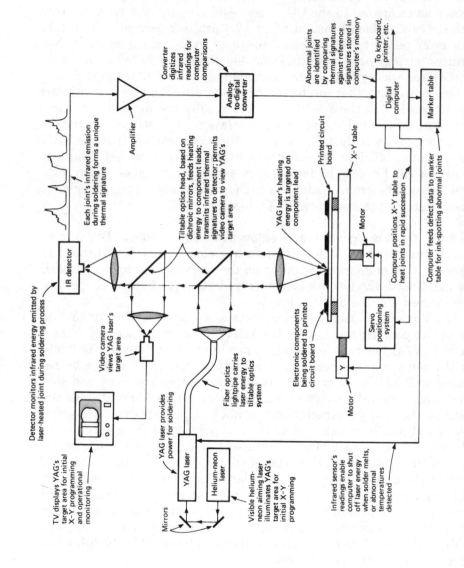

Figure 10.33 Optical system construction for a focused-beam laser soldering system. The system provides a condensed spot of radiation with various diameters. (*Courtesy Vanzetti, Inc.*)

Detector monitors infrared energy emitted by laser-heated joint during soldering process

Each joint's infrared emission during soldering forms a unique thermal signature

Amplifier

Converter digitizes infrared readings for computer comparisons

Analog-to-digital converter

Abnormal joints are identified by comparing thermal signatures against reference signatures stored in computer's memory

To keyboard, printer, etc.

Digital computer

Marker table

IR detector

Tiltable optics head, based on dichroic mirrors, feeds heating energy to component leads; transmits infrared thermal signatures to detector; permits video camera to view YAG's target area

YAG laser's heating energy is targeted on component lead

Printed circuit board

X-Y table

Motor

Computer positions X-Y table to heat joints in rapid succession

Computer feeds defect data to marker table for ink-spotting abnormal joints

Video camera views YAG laser's target area

Electronic components being soldered to printed circuit board

Servo positioning system

Fiber optics lightpipe carries laser energy to tiltable optics system

Motor

TV displays YAG's target area for initial X-Y programming and operational monitoring

YAG laser provides power for soldering

YAG laser

Helium-neon laser

Mirrors

Visible helium-neon aiming laser illuminates YAG's target area for initial X-Y programming

Infrared sensor's readings enable computer to shut off laser energy when solder melts, or abnormal temperatures detected

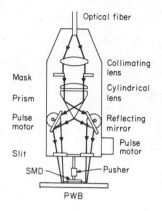

Optical fiber

Collimating lens

Mask

Cylindrical lens

Prism

Pulse motor

Reflecting mirror

Pulse motor

Slit

SMD

Pusher

PWB

Figure 10.34 Optical system construction for a diffused beam laser soldering system. The diffused beam is shaped in a triangular configuration in order to obtain simultaneous reflow.

rotated 90° in order to reflow the remaining two sides. Furthermore, during the reflow process, the component is mechanically forced against the termination pads by what is called a "pusher" so that all leads are in contact with the lands.

Bibliography

Capillo, Carmen, "Hot Bar Reflow Soldering of Ribbon Leaded Devices," *Soldering Technology Seminar Proceedings,* Naval Weapons Center, China Lake, California, February 1981.

Capillo, Carmen, "The Assembly of Leadless Chip Carriers to Printed Wiring Boards," *IPC Fall Meeting Proceedings,* October 1982.

Flattery, David, "Belt Furnaces for Microelectronics Processing Applications," *Microelectronic Manufacturing and Testing,* February 1983.

Hyman, H., and Peck, D. J., "Vapor Phase Reflow for Surface Mounting," *Soldering Technology Seminar Proceedings,* Naval Weapons Center, China Lake, California, February 1983.

Johns, W. E., and Lea, C., "Liquid Phase Soldering: Heat Transfer," *Brazing and Soldering,* no. 12, Spring 1987.

Lish, Earl F., "LASER Soldering of Surface Mounted Connectors—the Importance of Lead Configuration, Finish, and Solderability," *Soldering Technology Seminar Proceedings,* Naval Weapons Center, China Lake, California, February 1983.

Ohino, Keiji, et al., "YAG LASER Soldering for Fine Pitch Quad Flat Package (QFP)," *IEEE Proceedings,* 1986.

11

Cleaning Techniques and Equipment

11.1 Causes of Cleaning Difficulties

11.1.1 Component types

SMAs populated with complex components such as leadless chip carriers, SOICs, and PLCCs present a geometry that can be very resistive to the cleaning solvent's penetration and replacement during the postsoldering cleaning process. The difficulty of removing flux residues from SMAs increases dramatically as the component off-contact distance (the air gap underneath the SMD) decreases. This difficulty is further increased when the surface area of the SMD increases and when the centerline spacings between the leads increase, especially when leads are on four sides of the SMD.[1]

Figure 11.1 illustrates typical SMDs, their off-contact distances, and the surface area covering the substrate surface. Leadless SMDs such as LCCs, chip resistors, and chip capacitors have no off-contact distance other that what is gained by the solder and pad thickness upon which the SMD is placed. In most cases with leadless SMDs the off-contact distance can range from 0.001 to 0 .005 in. When solder masks are used, this distance can even be less. For these reasons, it is recommended that LCCs should only be soldered with mildly activated fluxes.

On the other hand, even though chip resistors and capacitors have no off-contact distance, as described, contamination removal under these components can be very easy, even with conventional cleaning processes and more activated fluxes. The main reason for this is that these components have small surface areas and exposed passageways due to the minimum solder contact points which can cause a reduction in solvent penetration and escape. Furthermore, when wave soldered

SOT-23 (small outline transistors/diodes)

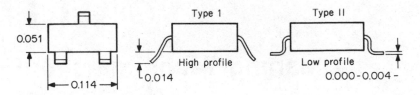

SOIC-16 (small outline integrated circuits)

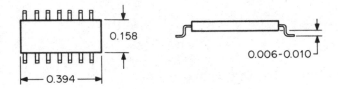

Quad-pack-68 (JEDEC proposed)

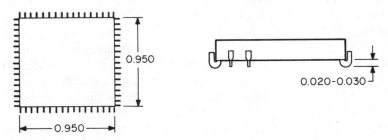

Figure 11.1 Typical surface mount components, their off-contact distances, and surface area covering the substrate surface.[1] Dimensions in inches.

and sometimes when other reflow processes are used, these components are bonded to the substrate with adhesives that act as a mechanical barrier to flux penetration under the component.

11.1.2 PCB design

PCB design and component layout can complicate the task of cleaning SMAs. Cleaning difficulties and field failures can result if the PCB design does not adequately address these potential problems. The following considerations should be investigated when designing a PCB for ease of flux or contamination removal:

1. Plated through holes that are placed underneath component bodies will allow flux to travel to the top side of the substrate if wave soldering is used. To prevent additional flux residues from traveling underneath SMDs, a solder mask should be used to cover the plated through holes on at least one side of the substrate.

2. Substrate cutouts should be avoided or kept to a minimum. When used, a breakaway substrate piece can reduce the possibility of solder or flux flow to the top side of the assembly during wave soldering.

3. Proper board thickness should be used in relation to board width or size so that stiffeners will not be required for wave soldering. Flux will become trapped between the stiffener and substrate, leaving flux residues on the substrate after cleaning unless they are removed prior to cleaning.

4. Solder masks that can maintain good adhesion without microcracking or wrinkling after several reflow soldering processes should be used. Photoimagable liquid solder masks have demonstrated their better temperature resistance and performance than dry film solder masks.

11.1.3 Component layout

Component layout, still another factor which can affect the difficulty of cleaning, is one of the most important concerns and often never a consideration by designers, which must be taken into account during the preliminary stages of design.[1] Two major parameters relating to component layout, which can contribute to the cleanability of the assembly, are lead extension directions and component orientation. These parameters have a great effect on solvent flow uniformity, velocity, and turbulence, both underneath the component and over the assembly, while the assembly is processed through the cleaner.

Typically the assembly travels on a horizontal conveyor belt into the solvent cleaner's throat or entrance and downward on a slope of 8 to 12°. During the last 18 in. on the downward slope of the conveyor, solvent spray typically hits the assembly at a nonperpendicular angle when the most effective machines are used.

At this point, the bulk of the flux residues are removed from the assembly and from underneath the components, since at this angle the solvent can travel underneath the components more easily if minimal obstruction does not occur. The same effect takes place when the assembly travels through the last sprays on the upward exiting part of the conveyor.

To take advantage of solvent cleaners designed in this manner, the lead extensions on components such as SOICs, chip resistors, and chip

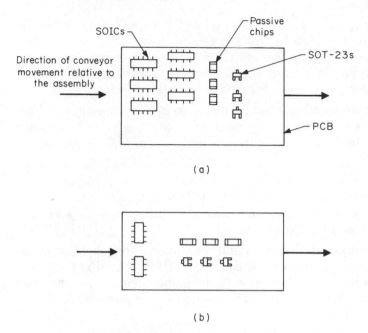

Figure 11.2 Component orientations on the substrate as they affect the degree of cleaning difficulty. (*a*) Preferred; (*b*) not preferred.[1]

capacitors should be positioned perpendicular to the forward travel of the assembly, as illustrated on Fig. 11.2. In this orientation, the downward flow of solvent across the board is not as easily blocked or deflected from passing underneath the component bodies, which, as mentioned, is the most difficult area to clean effectively.

Other factors that can contribute to cleaning difficulties are the type of solvent, machine, and process being used. These are discussed individually in their respective sections; however, there are still other areas that can substantially increase the level of cleaning difficulties.

11.1.4 Flux type

At the level being addressed, the major contributor to cleanliness difficulties is the type of flux used in the soldering process. Simply stated, as the percentage of solids and the activity of the flux increase, the more difficult it will be to achieve the same cleanliness levels reached by fluxes containing fewer solids and less activity. For instance, if the cleaning process is kept constant, i.e., the same solvent, machine, and process are used for cleaning assemblies with active and mildly active fluxes, it will be next to impossible for the assemblies soldered with active fluxes to reach the same or higher cleanliness levels than other assemblies soldered with milder fluxes. Therefore, depending on the effectiveness of the cleaning process and type of SMA, certain restrictions regarding the type of flux to be used for soldering may be necessary when acceptable cleanliness levels have been determined. Furthermore, if more activated fluxes are considered acceptable and the process has proved that it is capable of removing these fluxes to acceptable cleanliness levels, one still has to be concerned about the consequences with regard to life-support assemblies, which could fail due to a slight interruption of the cleaning process effectiveness. This is why mildly activated rosin (RMA) or rosin (R) fluxes are the only acceptable fluxes for use on military and aerospace assemblies. Under most conditions, even if these mild fluxes were left on the assembly, their noncorrosiveness and dielectric behavior under certain temperature and moisture limits would allow these fluxes to act as good electrical insulators.

11.1.5 Assembly postsoldering dwell time

Another major factor contributing to cleaning difficulties, and one that is often overlooked, is the time it takes the assemblies to reach the cleaning process after being soldered. This so-called in-process dwell time, which soldered assemblies may commonly experience in some manufacturing processes, could allow the flux to harden to a point where no cleaning process will remove it. In many cases, when flux is allowed to harden on the surface of an assembly, only mechanical removal methods may be effective, however impractical. Therefore, one must determine the maximum dwell time the assembly is allowed to experience prior to cleaning. This dwell time varies depending on the manufacturing process and type of flux; however, in any case, cleaning the assembly while it is still warm to the touch is preferred. Moreover, care should be taken not to expose assemblies to the cleaning solvent when they are too hot to handle since this can cause thermal shock or lack of solvent vapor condensation on the assemblies. As the assembly postsoldering dwell time increases, it should be expected that the assembly will require a significant increase in cleaning exposure time. It is best to minimize assembly postsoldering dwell times.

11.1.6 Reflow process

Finally, the reflow process itself and how it is used can either increase or decrease the level of cleanliness difficulty. No matter what the reflow process, at elevated temperatures, the temperature and dwell time are the conditions that also affect cleanliness difficulty. In whatever reflow process, one should not try to maximize preheat and reflow temperatures but install a setup which provides acceptable soldering yields. The reason for this is to prevent excessive flux deterioration to the point where significant compositional change will prevent effective or complete removal during the cleaning process.

11.2 Contamination

Cleaning is a process by which contaminants are removed. The primary cleaning task with SMAs is decontaminating flux residues after the reflow process; therefore, understanding the chemical constituents of postsoldering flux residues, which are the major assembly contaminants, is most important if one is to succeed in effective decontamination. Furthermore, when cleaning problems do result, classifying the type of contaminant and its potential source will be the major task in order to prevent its recurrence on the assemblies. Before jumping into specific information on contaminants, let us review some of the basic principles of contamination.

Contamination or contaminant, as defined by C. J. Tautscher, an authority on this subject matter, is "any surface deposit, or inclusion or occlusion and absorbed or absorbed matter, which is known to degrade to unacceptable level, the chemical, physical or electrical properties of the substrate or assembly." In review of this definition, no matter where it occurs on the assembly, a contaminant is a substance that will degrade or reduce the assembly's performance to an unacceptable level. Remember, performance may not be what is expected in the field but rather conformance to performance requirements as specified in a standard such as MIL-STD-810.

The contaminant's location, as described, can be on the surface, adsorbed, or absorbed. Contaminants located on the assembly's surfaces may be considered the most common. However, although located on the surface, the contaminant is usually adsorbed, absorbed, or both. Adsorption takes place when a contaminant can chemically react, on a molecular level, with the material with which it is in contact. An example of contamination that is adsorbed is seen when the solder mask is insufficiently cured. The uncured or non-cross-linked portions of the solder mask material can be considered "available" or unbonded material which can react with a contaminant such as flux residues, especially during the reflow process when heat is applied. When the reaction takes place, the contaminant or contaminants within the flux residues will have then been adsorbed. Their removal, which is sometimes impossible without damaging the assembly, can only be accomplished by breaking the weak van der Waals covalent bonds which hold the contaminant to the solder mask. If not too extensive, this type of adsorbed contaminant may be chemically removed by using a high boiling point (bp) solvent such as tetrachlorodifluoroethane (bp = 185°F) or 1,1,1-trichloroethane (bp = 164.5°F), which may soften and react with the contaminated solder mask and affect removal via breaking the weak covalent bonds.

Absorbed contamination penetrates the porous areas of the assembly. In fact, to a certain extent, every material is porous or rough where the mobile contaminant can be held more tightly to the surface of the material via mechanical anchoring. Fluxes, which start as fluids, certainly can easily find these porous areas and, by capillary action, can penetrate below the materials surface and, upon cooling and solidification, have a very strong mechanical hold.

As also mentioned, contaminants can be an "inclusion" or an "occlusion." Inclusions are typically solid particles embedded in a material such as solder mask or a plated deposit that protrudes beyond the material's surface. Occlusion contaminants can be the same type of solid particles; however, they are actually encapsulated within the material they are contaminating.

In most cases, inclusions and occlusions originate during the fabrication process of the PCB due to inadequate particulate matter control or removal prior to the next process step. Certainly the cleanliness and proper contamination control of PCB laminates is a major concern; however, it is beyond the scope of this section and is therefore recommended that the reader review the literature dedicated to that level of cleaning.

11.2.1 Types of contaminants

The types of contaminants on PCB assemblies that can produce electrical and mechanical failures over short or extended periods of time can be classified as shown in Fig. 11.3. To generalize, the several different types of contaminants can be categorized as either polar or nonpolar. Polar contaminants have eccentric electron distribution;

Figure 11.3   Types of Contaminants and Their Possible Origins

Contaminant type	Origin
Organic compounds	Fluxes, solder masks, tape, fingerprints
Inorganic insoluble	Photoresists, PCB processing, flux residues
Organic metallic	Flux residues, white residues
Soluble inorganic	Flux residues, white residues, acids, water
Particular matter	Airborne matter, debris

that is, the electrons that act as the "bond" between the atoms which make up the material are shared unevenly. This uneven sharing of electrons in a molecule such as HCl is produced from the different electronegativities of the atoms. With HCl, the Cl atom has the higher electronegativity, and, therefore, the pair of electrons shared by both atoms associates more with the Cl atom. This behavior in a molecule is identified as a "polar" characteristic. When disassociation of polar molecules such as HCl or NaCl occurs, as shown below,

$$NaCl + H_2O \rightarrow Na^+ + Cl^-$$

$$HCl + H_2O \rightarrow H_3O^+ + Cl^-$$

positive and negative ions are produced. Typically, these "free" ions are good conductors and, therefore, very capable of causing circuit failures. Also, these ions usually react strongly with metals; the resulting corrosion can lead to field failures of electronic systems. On the other hand, polar contaminants can also be nonionizable, which means the positively and negatively charged atoms will not disassociate. However, when nonionizable polar contaminants are presented in an electric field, during elevated temperatures, or with other types of stresses, the atoms of different electronegativities can align themselves so that an electric current can be carried.

Nonpolar contaminants are compounds that do not have an eccentric electron distribution and will not disassociate into ions or carry an electric current. These types of contaminants are mostly made of long-chain hydrocarbons or esters of fatty acids consisting of 20 or more carbon atoms. Nonpolar contaminants are generally good insulators with high dielectric strengths. Well, why then are they considered contaminants? These nonpolar contaminants can certainly prevent effective circuitry testing simply by acting as a dielectric between the test probe and probe contact on the PCB. Furthermore, it is possible for ionic contaminants to be held within the nonpolar contaminant or covered by them and later be freed or exposed to allow potential electrical failure. In most cases, nonpolar contaminants will not produce corrosion or electrical failures; however, they may prevent solderability and solder mask adhesion along with the other concerns mentioned above.

11.2.2 Flux residues

Since fluxes are necessary for the soldering process, i.e., for the most part they are used to remove oxides from the metals to be joined and to reduce the surface tension of solder, it is safe to assume that they are also the major source of contamination on the assembly. Therefore, if one is to effectively remove flux from the assembly, it is important to understand the chemistry of the postsoldering residues that are the reactionary products of heat-modified flux.

In order to accomplish this, it would be useful to know the chemistry or formulation of the flux being used. This, however, presents a major problem in that exact formulations are not usually made known to the user. Furthermore, flux manufacturers may provide as many as 30 standard flux formulations with significant variations in their chemistry and behavior during and after soldering. In addition to these standard formulations, nonstandard flux formulations that are supplied to companies with special requirements may increase the total number of formulations beyond 100. Fortunately many of the available flux formulations can be grouped into two major categories: organic solvent–soluble and water-soluble. The organic solvent–soluble fluxes are the most common and are likely to become the standard flux type for SMAs, whereas the water-soluble types have been the standard flux for conventional through-hole assemblies.

Since the organic solvent–soluble fluxes have been extensively used in most solder pastes and are widely used for the dual-wave soldering process, our discussions regarding flux residues will mainly be directed to these fluxes. However, if the water-soluble fluxes are being used on any one of the various types of SMAs, it is important to characterize the remaining flux residues on the assemblies. In all cases, whether using organic solvent–soluble or water-soluble fluxes, it should be understood that fluxes which are processed through typical surface mount reflow processes can leave residues which are different in composition and more difficult to remove, even if mild formulations are used. This is mainly due to longer heat exposure times when compared to conventional reflow soldering techniques.

The organic solvent–soluble fluxes can contain either gum rosin, synthetic rosin, or resin along with solvents, wetting agents, and activators. Unfortunately, at this time qualitative and quantitative postsoldering analyses of fluxes containing synthetic activated rosins have not been performed. However, the popular rosin-based flux formulations used in solder pastes and some new surface mount rosin dual-wave formulations have been analyzed so that there is a better understanding of "what" the change is in the chemistry of the flux after the exposure to the reflow process.

A rosin flux is a solution of rosin in a suitable solvent. Typically, the solvent is an alcohol such as isopropanol, glycol ether, or a terpene. In most cases the solvent is evaporated off during preheat and reflow. Only rarely do solvents become a part of the flux residue.

Wetting agents may be incorporated in a liquid flux to produce desirable characteristics such as an even foam head, smooth solvent evaporation, and uniform coating of the PCB. These proprietary chemicals can be any of a great variety of chemical substances but are usually used in very small concentrations in the rosin fluxes and are, therefore, inconsequential to the problem of residue removal.

Rosin, a natural resin extracted from the oleoresin of pine trees and refined via steam and vacuum distillation, is 90 percent resin acids, some organic fatty acids, and terpene hydrocarbons. Available as gum, wood, and tall rosins and sometimes chemically modified, the most widely used type for liquid and paste fluxes is the water white (WW) gum rosin grade and, with tall oil rosin, a common grade for use in solder pastes.

The resin acids are divided into the abietic and pimaric types, together accounting for over a dozen different acids. Both types are tricyclic monocarboxylic acids usually containing two double bonds (centers of unsaturation). Only in the abietic-type acids are the double bonds conjugated. This accounts for the instability of the abietic-type resins in the presence of heat, mineral acids, and air oxidation.

The oxygen in the air readily reacts with the conjugated double bonds and associated active sites of the abietic-type acids to produce ketones, alcohols, and diols through either a peroxide or epoxide mechanism. Heat, of course, accelerates the oxidation. Some pimaric-type acids may also contain active sites for the absorption of oxygen. Likely areas for oxygen pickup are during rosin and flux manufacturing, foam fluxing, bakeout (solder pastes), preheating, and soldering. Furthermore, oxidized rosin is less soluble in many organic solvents, is a less effective flux, and may become polymerized. Therefore, to minimize the problems associated with oxidized rosin, it is imperative to obtain a rosin flux from a proven source, to not overheat it during preheat or bakeout and soldering operations, and to change it periodically when using it in liquid form such as in a foam fluxer. Also, when solder pastes are used, the best way to prevent oxidation to both the metal powder and rosin is to mix and store the material under dry nitrogen. This will significantly prolong the shelf life, minimize the formation of solder balls, and minimize the oxidation of the flux which, in any case, could leave postsoldering residues which are more difficult to remove.

As in any other type, the largest chemical changes in rosin-based fluxes take place during the reflow process. It was once thought that the major chemical change in a rosin flux upon soldering was polymerization. Polymerization is the chemical reaction in which two or more monomers or polymers of the same kind are united to form a molecule of a higher molecular weight. Although polymerization of rosin can occur upon ultraviolet light or thermal exposure, it has been found that isomerization is the dominating reaction of rosin-based

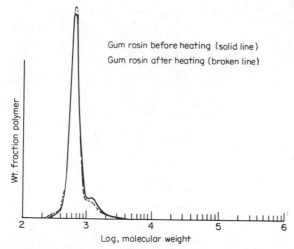

Figure 11.4 Chromatogram of gum rosin before and after soldering showing no change in molecular weight; therefore, mostly isomerization occurs. (*Courtesy Dow Chemical.*)

fluxes which have been soldered. In this case, isomerization is an atomic rearrangement of rosin without an increase in molecular weight; hence, no cross-linking occurs as does in polymerization, but there are changes in properties such as solubility in an organic solvent. Figure 11.4 demonstrates that there is no shift in the molecular weight of gum rosin before and after reflow soldering when measured using a gel permeation chromatograph, which proves that isomers of gum rosin make up the organic portion of the flux residues. Further analysis, as shown in Fig. 11.5, demonstrates that there are at least three major organic rosin flux residues left on the assembly after soldering consisting of abietic and dehydroabietic acid and neoabietic acid. Since gum rosin consists primarily of abietic acid, the isomerization of this acid under moderate heat can produce neoabietic acid and under high heat can produce dehydroabietic acid, as shown in Fig. 11.6. On postsoldered assemblies, other isomers have been detected such as dihydroabietic, tetrahydroabietic, and pyroabietic acid, which are isomers of abietic acid along with pimaric acid (a major portion of gum rosin like abietic acid) and isopimaric acid, an isomer of pimaric acid.[2]

Up to now, the major organic materials present in rosin flux residues have been discussed; however, the nature and reactions of the ionic flux residues have not yet been described. Ionic residues by far are the most damaging to the performance of an electronic assembly since, as mentioned, they can be corrosive and electrically conductive.

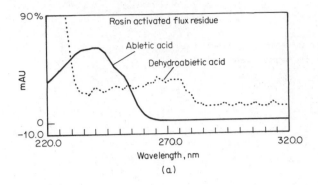

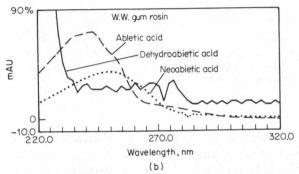

Figure 11.5 (*a*) Ultraviolet spectra of rosin acids in rosin-activated flux residue after wave soldering; (*b*) rosin acids present in water white gum rosin. (*Courtesy Allied Chemical.*)

Although called acids, the organic acids (abietic and pimaric acids) in a rosin flux are very mild and by themselves are not chemically active enough to reduce metal oxides to promote rapid solder wetting conditions. To increase the ability of rosin fluxes to clean oxide metal surfaces, activating agents are added to produce RMA (rosin mildly activated) and RA (rosin activated) fluxes. The less aggressive, noncorrosive RMA fluxes, which are used for high-reliability electronics, are typically formulated with activators consisting of organically bound halogen compounds. Because RMA fluxes must be noncorrosive as specified in the military specification Mil-F-14256, only a limited amount of these activators may be added. Therefore, rosin and its by-products are the primary and most important materials in the flux residue. Therefore, RMA flux residues, as discussed, consist primarily of rosin organic acid isomers.

Regardless of the soldering process chosen and since the solderability of substrates and components is a major task, especially

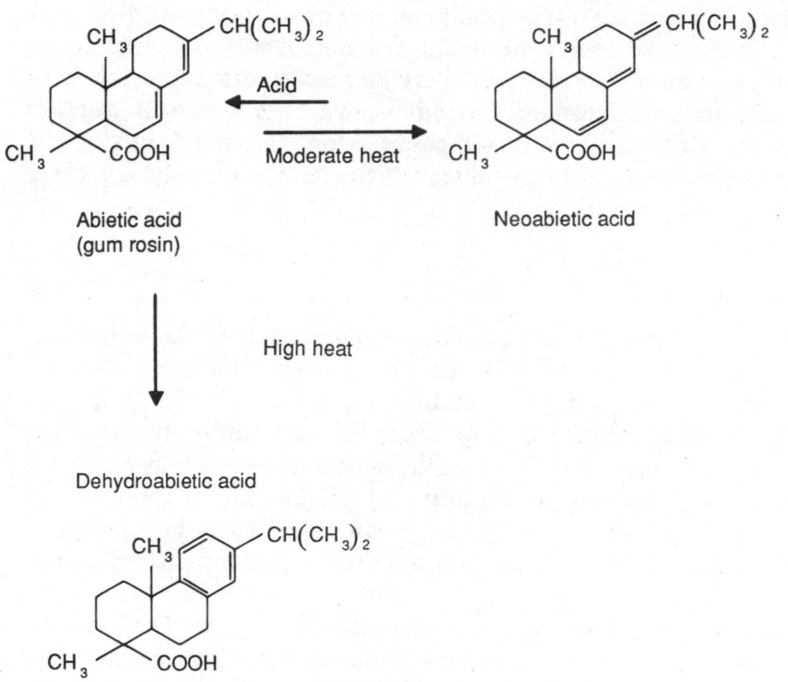

Figure 11.6 Transformations of abietic acid upon reflow soldering into related isomers or flux residues.

in high-volume products, faster soldering or wetting speeds require more activation in the soldering flux. To accomplish this, activators such as amine hydrohalides and alkanol amine hydrohalides are added to rosin flux to place them in the RA category as specified in Mil-F-14256. When heated, these activators can release the hydrohalide (HCl or HBr) through disassociation as shown in the following equation:

$$R_2^+NHCl^- \rightarrow R_2NH + HCl$$

This strong mineral acid, HCl, easily reacts with the oxide layer of the metal, facilitating its removal and exposing pure metal to the solder:

$$CuO + 2HCl \rightarrow CuCl + H_2O$$
$$\text{(green)}$$

The green material frequently found when using rosin fluxes on copper is due to the formation of copper rosinates, or CuCl, if activated fluxes are used. The copper salts so formed are dispersed in the flux residue and are easily and completely removed by those cleaning solvents which contain polar components. These green residues are typically thought of as corrosive products; however, many studies have shown that most corrosive products are not green, whereas many green residues are noncorrosive and are just a cosmetic condition. This does not mean that "cosmetic" residues, whether green or another color, are acceptable since it is still possible for these noncorrosive residues to mask over potential residues which could lead to the corrosion cycle and corrosive residues.

11.2.3 Insoluble residues

Insoluble residues, which are sometimes reported as white or tan residues, can be found on assemblies that have been soldered with rosin-based fluxes. These insoluble residues have typically been demonstrated to be only a cosmetic condition and are different from the white lead carbonate ($PbCO_3$) residue as mentioned in Sec. 11.2.4. These insoluble residues are usually not visible to the unaided eye and are only seen after the solvent cleaning process. Furthermore, they seem only to form with rosin-based fluxes and not with synthetic activated fluxes.

Early analysis of these insoluble residue was thought to be a reaction between the solvent, typically a fluorocarbon-alcohol azeotropic blend, and heat-modified rosin. Furthermore, solvent-extracted resistivities of these insoluble residues showed that they are non-soluble in water and only partially soluble in alcohol, leading to the interpretation that they are nonionic. However, the insoluble residue's solubility in strong polar amides and in solvents containing N^- complexing agents suggests that this material does possess ionic character even though it is insoluble in hot or cold water.

Contrary to earlier studies, more recent experiments performed by D. G. Lovering[3] have provided substantial help in clarifying the cause of the formation and the identity of this insoluble residue. Upon solubility analysis of the insoluble residue, it proved to be partly insoluble in acetonitrile and tetrahydrofuran, but partly soluble in ethanol and acetone. From this behavior with ethanol, one may deduce that when it is so-called tan residue or has a tan color, the residue is mainly composed of unchanged rosin. However, when white in appearance after being used during the cleaning process, it can be expected that the alcohol in the azeotropic solvent has removed the rosin which has a certain amount of this insoluble residue dispersed in it. With the removal of the rosin, the nearly "pure" form of the insoluble residue appears white in color.

Electron-microscopic analysis of the white residue, as shown in Fig. 11.7, shows L-series lines for tin and chlorine with minor components of lead, copper, iron, and bromine. Obviously, tin is the major component of the white, insoluble residue and must be the product of a re-

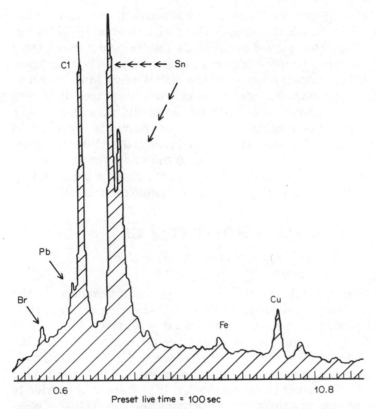

Figure 11.7 Link Systems 860 microprobe analyzer spectrum of the L-series tin lines obtained from the white residues.[3]

action between rosin and the molten tin-lead solder during the soldering process. The lead in the solder appears to react less than the tin; however, it is not totally inert during the soldering process. The copper and iron, which are always present in low quantities, should originate from the PCB and component leads, respectively. The chlorine and bromine originate from the activators used in the flux formulation.

With this analysis, it can be concluded that these insoluble white or tan residues are tin abiate salts caused by the reaction between tin-lead solder and rosin during the soldering process. The tan residues are simply undissolved rosin with dispersed amounts of tin abiate salts.

11.2.4 Corrosion cycle

When activated fluxes are used, whether solvent-soluble or water-soluble, it is essential that they be removed from the assembly soon after the soldering process, because these active fluxes can reduce strong mineral acids, which not only react with the metal oxides but also readily attack the cleaned metal leads and the solder itself. Once this reaction occurs, it produces metal halogen salts (when halogens are used in the activator), which combine with the flux binder such as rosin. When the time from soldering to cleaning increases, it is very possible that the salts can be left behind, especially when a nonpolar or semipolar solvent is used. In the presence of humidity, the exposed halogen salt residues become very conductive. Furthermore, the halogen salts (Cl^- or Br^-) will easily react with the solder, producing lead chloride ($PbCl_2$) or lead bromide ($PbBr_2$), which starts a perpetuating process called the *corrosion cycle*. The corrosion cycle itself can be shown as follows:

$$PbCl_2 + CO_2 + H_2O \rightarrow PbCO_3 + 2HCl$$

or
$$PbBr_2 + CO_2 + H_2O \rightarrow PbCO_3 + 2HBr$$
(humidity) (white residue)

As shown above, $PbCl_2$ reacts with CO_2 in the atmosphere, in the presence of humidity, and produces $PbCO_3$ (lead carbonate) and HCl. Once HCl is produced, it immediately reacts with the solder to produce more $PbCl_2$, therefore perpetuating the corrosion cycle. During the course of this cycle, large amounts of white $PbCO_3$ are formed, which can cover the areas on the solder where the reaction is taking place. Since $PbCO_3$ is not soluble in water, it can act as a barrier to the cleaning process in preventing or terminating the corrosive reaction. This is the main reason for cleaning assemblies soon after they have been soldered with an active flux.

11.3 The Theory of Cleaning

In order to understand how contaminants are cleaned from surfaces, it is important to understand how contaminants adhere to these surfaces. It must first be assumed that if a contaminant is adhering to a surface, there must be a bonding mechanism between the two. Otherwise, if there is no bonding between the contaminant and surface, it can be assumed that the contaminant will separate with little or no force. Contamination which is not "bonded" to the surface will usually present little difficulty in removing it.

Contaminants such as flux residues, solder balls, and tape residues normally have several bonding mechanisms between the contaminant and the substrate. Physical and chemical bonds are the so-called bonding mechanisms which cause contamination to adhere to the substrate. In most cases, both occur simultaneously, and the absolute distinction between the two may not always be agreed upon among scientists. For one to effectively remove these contaminants, it is necessary to first weaken and destroy the bonding mechanisms holding them together. In the following sections, let us gain an understanding of these bonding mechanisms and what it takes to destroy them when using cleaning solvents.

11.3.1 Physical bonds

Physical bonds that cause contamination adhesion can consist of a mechanical force and an absorption or capillary force which holds onto the contaminant at the substrate's surface. Contamination which adheres mechanically occurs due to the anchoring of the contaminant to the substrate's microscopic surface irregularities. Where copper has been etched away, the substrate itself provides a very irregular surface. It is estimated that the true surface area of an etched substrate can be approximately 50 times that of the apparent surface area. This is an ideal condition to promote and maintain strong physical "mechanical" bonds between the contaminant and substrate surface. Although not microscopic like substrate surface irregularities, mechanical bonds can also take place in areas such as between fine pitch leaded components, in plated through holes, and under a low-profile component. This "visible" mechanical bonding typically occurs only with a large volume of contamination such as flux, which can mold itself between these areas solidifying, and thereby locking, itself in.

As mentioned, absorption of contamination is the physical bonding mechanism that contributes to the adhesion of a contaminant. Absorption of a contaminant takes place when capillary forces draw the contaminants into a porous region of the substrate or assembly. Such absorbant substrates are more frequent than is commonly believed to contribute to the cause of contamination remaining on the substrate. For example, if the resin in the board laminate or even the resin in the solder mask is not mixed or cured properly, insufficient polymerization of the resin results. This leaves a homogeneous surface, but one that is chemically tender and reactive. Under the right conditions, contaminants such as flux residues could chemically bond to the unpolymerized resin molecules during the soldering process; however, if this does not occur, then during the solvent cleaning process the unpolymerized resin will dissolve away. Upon doing so, a very porous surface can occur, allowing contaminants to be drawn within the porous surface via capillary forces.

Another, more common instance of contaminants entrapped via capillary forces is when solder mask delaminates from the surface of the substrate. Since SMAs can undergo multiple reflow soldering steps, the loss of solder mask adhesion, particularly for dry film types, is very likely. Upon the occurrence of solder mask wrinkling or lifting near its edges, small cracks or spaces will ultimately produce an opening where liquid contaminants can be drawn far underneath the solder mask. In this case, removal of the contaminant is impossible without destructive measures.

11.3.2 Chemical bonds

Contaminants, or any combination of materials that are chemically bonded together, are formed either by valence bonding or by a phenomenon called adsorption. These two types of chemical bonds can both easily occur during the soldering process between the flux residues and substrate materials. With few exceptions, contamination adhering to the substrate dominated by chemical bonds instead of physical bonds is by far the most difficult to remove and, in some cases, next to impossible to remove to levels of ineffective quantities. In all cases, steps should be taken by process and quality engineering to prevent or minimize chemical bonding from occurring.

Valence bonding is the only true chemical bond, and it occurs when a chemical reaction takes place at the meeting point of two or more substances. These substances can be liquid, gaseous, or solid. Valence bonding occurs at the atomic level, when two or more atoms bond together by sharing electrons in their outermost shell, called the valence shell and, hence, the designation valence bonding. The oxidation of metals such as component leads and metal copper foils common to PCBs can be given as examples where valence bonding occurs. Once exposed to the atmosphere, oxygen can chemically react with these exposed metals to provide a thin outer layer of metal oxide. Since this is unavoidable, flux, which later becomes a contaminant, is used to chemically react with the metal oxides in order to reduce or break the valence bonds between the metal and oxygen, leaving a clean, unoxidized metal surface. In some instances, due to limited process time, excessive preheat, or flux activation, heavy metal oxides can build up to a point where they cannot be removed under normal conditions, which leads to solderability problems.

Contaminants that adhere themselves predominantly by adsorption are yet another concern prior to and during the cleaning operation. Adsorption, a phenomenon that is still not clearly understood, occurs at a molecular level and is the key to effective adhesion of any substance onto a nonabsorbent surface. At the molecular level and always within the near surface of the substrate, weak molecular forces, sometimes called *van der Waals forces,* can cause an attraction affinity for a contacting material. This kind of adsorption is a physical one usually called wetting.

Adsorption can also occur via an enhanced form of a chemical reaction and is more commonly called chemisorption. A well-known type of chemisorption is the formation of *intermetallic compounds* when molten solder makes contact with a clean copper surface. Another possible form of chemisorption related to substrate contamination or the formation of "interorganic" compounds is with insufficiently cured substrate laminates and solder masks during the flux-soldering process. During the preheat stages of soldering, flux is capable of reacting with the unpolymerized or uncross-linked hydrocarbons in the laminate or solder masks. This reaction can leave behind weakly bonded organic compounds on the surface of the insufficiently cured material. Often these organic compounds are white and can be removed by a lengthy soak and spray in high boiling point solvents such as 1,1,1-trichloroethane or tetrachlorodifluoroethane. Of course, with the right conditions, this later case of chemisorption could have also occurred via valence bonding, which would be more likely to cause the permanent adhesion of the organic compound.

11.3.3 The mechanisms of removing contaminants

Whether by physical or chemical bonding or both, once contaminants adhere to the PCB assembly, it becomes necessary to destroy these bonds so that their removal becomes possible. The destruction of bonds is an endothermic reaction, i.e., energy must be supplied to cause bond destruction. Since during the reflow process, the spreading of flux is essentially over all the assembly and underneath the SMDs, it is necessary to apply this energy everywhere. This energy applied in the form of a solvent and the process of its application can be accomplished by *dissolution* or *saponification* of the contaminant.

Dissolution of contaminants via various cleaning solvents has become widely used for flux removal from SMAs. This technique simply involves dissolving the flux residues in the solvent and repeating this process until all the contaminants are dissolved and driven away from the surface they are adhering to. An analogy of this behavior can be given when a teaspoon of honey is placed in a cup of warm tea. The honey will be considered the solute and the tea the solvent. When the process begins, the honey flows off the spoon much faster due to the increase in temperature supplied by the warm tea. Since honey has a high viscosity, stirring or agitation becomes necessary to completely remove it from the spoon. In much the same way, flux residues are removed from the assemblies they have contaminated; however, the geometries that are present, such as small clearances under SMDs, component leads, and component types, make the cleaning process

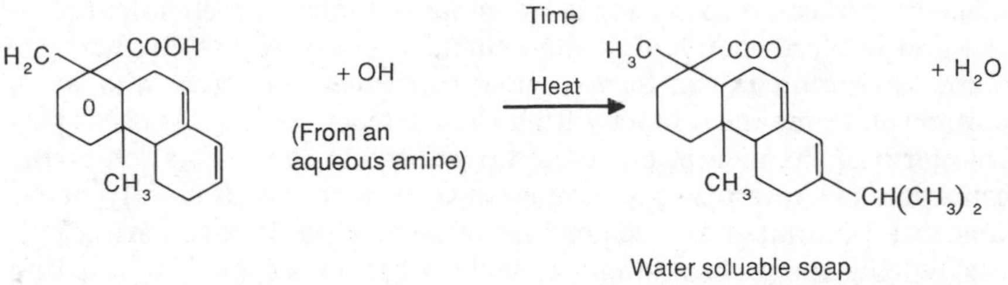

Figure 11.8 The saponification of abietic acid (rosin) with an aqueous amine which forms a water-soluble soap.

much more complicated than what appears on the surface. Therefore, it is of major importance to clearly understand the mechanical, physical, and chemical properties of the solvent and the process of solvent application in order to reach the desired cleaning levels within a limited period of time. These concerns, which are also related to saponification, are discussed in the following sections.

Saponification has been used as a decontamination method in removing rosin flux for a significant period of time for the purpose of using safer aqueous cleaning solutions instead of solvent base. Abietic and pimaric acid, as mentioned in Sec. 11.2.2, form a major part of rosin and are almost entirely insoluble in water. Because of this chemical behavior between water and rosin, a somewhat difficult problem occurs when one wants to use an aqueous cleaning process to remove rosin flux. To overcome this problem, surfactants, usually alkaline amines, are used with water; they react chemically with rosin to convert it to a water-soluble soap and, hence, produce a saponification reaction, which is shown in Fig. 11.8. Ammonium hydroxide and monoethanolamine are typical surfactants used in the saponification of rosin and are also very good wetting agents. Their ability to wet and stay wetted to assembly surfaces makes their complete removal somewhat difficult, especially in difficult-to-reach places. Because of this behavior, it is unlikely that rosin-based fluxes will be best removed via this technique when cleaning SMAs. In any case, if saponification is used, vigorous high-pressure spraying and a final clean deionized water rinse should be mandatory final cleaning steps.

11.3.4 The physical, chemical, and mechanical behavior of solvents affecting decontamination

Unlike years ago, there are now dozens of new types of solvents, usually azeotropic mixtures or solvent blends of either chlorinated or chlorofluorinated hydrocarbons, with various alcohols and stabilizers. These various types, which are discussed in Sec. 11.3.5, have different physical and chemical characteristics and can be used via several mechanical application processes (agitation, spraying, immersion, and ultrasonics). When a choice has to be made by a potential user, using the advice from the solvent manufacturer or even someone who is actually using the solvent should not be the final step in deciding what solvent is best for you. In most cases the manufacturers of these solvents and even some of the users are very biased toward the solvent with which they are involved. The interesting part to this is that the solvent manufacturers all have very convincing marketing approaches that can easily sway the potential user. Therefore, in order to determine the best solvent for your needs, an understanding of the physical, chemical, and mechanical characteristics of solvents which affect efficient cleaning of SMAs is necessary. This alone is still not sufficient to make a final decision; however, it will give us a foundation of knowledge to work with so that a scientific approach can be applied to the advice given by others. Furthermore, we must also realize that a solvent which is the best for one person or company may not be the best for another. For example, testing may show that solvent A removes more ionic contamination than solvent B and therefore one may say solvent A is the best. This may not be true for some if solvent A is more expensive than solvent B and the additional amount of cleanliness given by solvent A makes no significant difference to the product's reliability. However, those who require the highest ionic cleanliness levels, regardless of costs, would consider solvent A the best. Let us now review the solvent characteristics that should play a part in our decision in choosing the solvent that best suits our needs.

11.3.4.1 Wetting and surface tension. In order for a solvent to dissolve and carry away contamination, it must first wet the substrate that is contaminated, spread, and wet to the contamination. If this does not occur effectively, then the solvent will not be as effective in contamination removal.

Wetting can be calculated from the contact angle (dihedral) formed by a drop of solvent at the point where it contacts the substrate. Experimentation has shown that nonwetting occurs when the contact angle (σ) is greater than 90°. If $0° < \sigma < 90°$, partial wetting exists. Obviously, the most desirable cleaning condition is one where the solvent spreads spontaneously over the substrate surface and where the contact angle is close to, or preferably at, 0°. But at 0° it is more precise to speak of a spreading coefficient:

$$\cos \sigma = \frac{\gamma_s - \gamma_s^1}{\gamma^1}$$

where γ_s = force energy of solid substrate
  γ^1 = surface tension of solvent
  σ = equilibrium contact angle formed by drop of solvent

If the coefficient is positive, the solvent wets the substrate easily. The problem with this procedure is that it is very difficult to determine the interfacial free energy of the solid interface. Both factors are required if an accurate value for the equilibrium contact angle is to be formed.

To overcome the problem of an unknown spreading coefficient, it is necessary to use the concept of the critical free energy of a solid. This experimentally devised property is used to predict whether complete wetting or spontaneous spreading will occur. If the surface tension of the solvent is less than the critical free energy of the solid, complete wetting will occur; but if the surface tension is greater, partial nonwetting can result. Values for clean solids are provided in Fig. 11.9a. The values for surface tensions of some solvents are listed in Fig. 11.9b. Based on these data, it can be observed that the solvents commonly used in cleaning will completely wet the common materials used as substrate or components. For example, since the critical free energy for epoxy glass is 38, solvents like 1,1,1-trichloroethane and trichlorotrifluoroethane will spread spontaneously over this material. However, water with a surface tension of 72.75 will nonwet or partially wet to epoxy glass since its surface tension is more than twice that of this critical surface free energy of glass epoxy. Based on this analysis, a condition of low surface tension will promote solvent wetting or spreading onto a substrate. A condition of higher surface ten-

Figure 11.9   (a) Critical Surface-Free Energies of Polymers; (b) Typical Surface Tensions at 68°F (20°C)

a. Polymer	γ_s (dyn/cm)
Polytetrafluoroethylene	18.5
Polypropylene	29
Polyethylene	31–31.5
Polystyrene	30–36
Polyvinyl chloride	39
Polycarbonate	29
Polyamide-epichlorohydrin resin	52
Urea-formaldehyde resin	61
Epoxy glass	38
b. Solvent	γ (dyn/cm)
Trichlorotrifluoroethane (fluorocarbon)	17.7
1,1,1-Trichloroethane (chlorocarbon)	25.1
Ethanol	22.3
Water	72.75

sion may result in insufficient or poor solvent wetting leading to solvent "beading."

11.3.4.2 Capillary action. Using a solvent with the best wetting capability does not ensure that contaminants will be effectively removed from under and between SMAs and from around the point of contact of some of the more complex SMAs such as connectors or sockets. Therefore, a solvent must also have the ability to easily penetrate these tight spaces, remove contamination, and exit the space so that the cycle can be repeated with clean solvent until the contamination is removed.

The movement of solvent into restricted spaces is analogous to normal capillary action. The driving force for capillary action is pressure differential existing across the meniscus. For any spherical capillary, the pressure differential calculated to predict solvent penetration is directly proportional to the surface tension. This does not mean that solvents (such as water) with high surface tensions are preferred due to their higher capillary action of traveling into tight places. This is due to the fact that even though strong capillary action is desirable, the solvent's capability of exiting from those tight areas becomes more difficult. It is of great importance to realize that in order to clean effectively under SMAs or in tight places a solvent must penetrate and exit easily so that circulation and solvent exchange can occur multiple times during the cleaning process. Figure 11.10 illustrates the rate of penetration of solvents into a horizontal capillary. Upon analysis of these data, water's capillary rate of penetration under SMAs would be expected to be very fast; however, due to its high surface tension, it may be difficult to remove or provide insufficient exchange rates of clean water. The fluorocarbon solvent blends have the lowest capillary rate of penetration and surface tension at operating temperatures, which may be considered the minimum "preferred" conditions. On the other hand, the chlorocarbon solvent blends can be considered to offer compromised physical properties of rate of penetration and exchange.

Figure 11.10 Rate of Penetration of Solvents into a Horizontal Capillary

Solvent	T, °C	$\gamma \cos \sigma / n$
Fluorocarbon blend	25	26.4
Chlorocarbon blend	25	31.4
Water	25	40.4
Fluorocarbon blend	40	28.0
Chlorocarbon blend	73	40.34
Water	70	112.7

11.3.4.3 Viscosity. Based on the aforementioned properties of solvents, a so-called universal solvent that would best clean a variety of SMAs would provide good wetting and have a low surface tension and high capillary forces to aid in its penetration under SMAs. If this same solvent has a high viscosity, its exchange rate from underneath SMAs would be very slow since more force would be required to push the solvent out. Therefore, a low viscosity would aid in the solvent's ability to accomplish multiple exchanges from underneath SMDs.

11.3.4.4 Density. The argument for including density in an evaluation of solvents centers around the effect of gravitational forces exerted on the solvent. When solvent vapors condense on the assembly during the cleaning process, gravity will cause the condensed solvent to flow downward. For assemblies positioned horizontally, the solvent tends to spread farther and more evenly with higher density solvents. Furthermore, reducing losses to the atmosphere, which can be a considerable factor in costs, would favor using higher density solvents.

11.3.4.5 Boiling point temperature. The effects of solvent temperature on cleaning performance are not usually taken into consideration. Cold cleaning can be employed adequately for isolated contaminated areas, but most solvents are used at or near their boiling point. Different blends boil at different temperatures, and some of the physical characteristics, as mentioned, are affected by the change in solvent operating temperature.

If vapor condensation is part of the cleaning cycle, a higher vapor temperature leads to a greater volume of condensate and condensate runoff since condensation ceases when the temperature of this assembly is equal to that of the vapor. An increase in the temperature of the solvent usually leads to an increase in the solubility of the contaminant and the amount of contaminant the solvent can hold in solution. Since solubility increases with an increase in solvent temperature, greater quantities of contamination can be removed in a shorter period of time. This condition is of major importance with an in-line conveyorized wave soldering and cleaning system where the cleaner conveyor must not move any slower than the wave solder conveyor.

11.3.4.6 Solvency. Relatively speaking and typically under SMAs, only a small amount of solvent is likely to contact the contaminant. Therefore, it is important to use a solvent with the highest solubility factor, especially when there is a limited amount of time to clean such as in an in-line conveyorized cleaning system. In all cases, the higher the solvency of the contamination, the less time required for its removal and the faster the cleaning cycle. A word of caution should be given, in that, solvents with high solubilities are more likely to attack certain polymers typically used as component standoffs and adhesives in bar code labels than solvents with lower solubilities.

When comparing solvents for solubility, there are several methods that can be used. One is to select a number of solvents, clean the assembly in each, and then measure the effectiveness. This is a time-consuming approach and still does not cover all the solvents currently available.

Another approach is to compare solvents based upon the results of a highly standardized test. However, it has been shown that these standard tests, especially those based on kauri-butanol, are inadequate for defluxing evaluation. The kauri-butanol test is generally used to describe the ability of a solvent to dissolve a gum rosin; however, scrutinizing this test shows that the solute kaurigum rosin is not the same as the rosin used in flux. In addition, this test favors solvents miscible in butanol.

In order to make fair solubility comparisons between different solvents, it is necessary to use the flux residues that are predominantly left on the assembly after the soldering and cleaning processes. Because rosin-based fluxes are used in most solder pastes and in dual-wave soldering, determining the characteristics of the flux residue with these types of fluxes is useful since this residue would be the best solute to use in a solubility comparison of various solvents.

Ultraviolet spectrophotometry analysis of postsoldering and cleaning rosin-based flux residues has been performed in order to determine their chemical properties.[4] Their analysis shows (Fig. 11.11) that abietic acid, the major ingredient of rosin, remains relatively un-

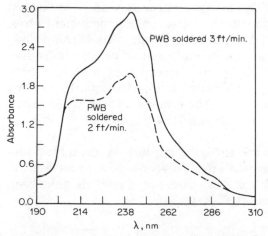

Figure 11.11 The ultraviolet spectra of practically unchanged rosin (abietic acid) flux residues left on an assembly after wave soldering and cleaning.

changed after being exposed to various wave soldering conditions. This would also be expected if vapor phase and IR reflow methods were chosen since their temperatures are normally lower than that used in wave soldering. As a result of this analysis, abietic acid has been used as a model compound for rosin flux residues remaining on an assembly subject to normal soldering conditions. In Fig. 11.12, the saturation solubility of abietic acid in 18 solvents and blends is shown. The plot demonstrates that the solubility of abietic acid increases as the solubility parameters of the solvent approach those of the acid. The data also show that chlorocarbon blends are superior to fluorocarbon blends with respect to rosin solubility except with the fluorocarbon–methylene chloride azeotrope. Unlike the other fluorocarbon solvents, this azeotrope has a high rosin solubility due to methylene chloride's very high solvency of rosin. This solvent is typically used for degreasing large metal parts and not for electronic assemblies and it can easily react with commonly used plastics and components.

11.3.4.7 The effects of solvent spraying. Static solvent baths or vapor zones alone are not enough to remove flux residues from soldered assemblies, regardless of the solvent used. It is mandatory to apply the solvent with some type of mechanical action that promotes the erosion of flux residues from the assembly. Spraying, ultrasonics, and boiling-spray immersion are the common types of mechanical action used to substantially increase the effectiveness of decontamination. Of these three types, spraying of solvents is the most commonly used and understood technique.

In most cleaning systems, solvent spraying is usually incorporated so that pressures from 5 to 25 psi are generated. Depending on several factors such as component-to-substrate clearances, component sizes, machine type, conveyor speeds, and solvent type, this range of low-velocity solvent spraying may not be sufficient to perform adequate decontamination. With difficult-to-clean assemblies, primarily those with component-to-substrate clearances of less than 0.010 in. and component lead counts of 44 and up, the removal of very small particles and flux residues from underneath the SMAs can become extremely difficult.

To understand the effects solvent spraying has on assemblies and under SMAs, one must review the effect solvent velocities have in regard to drag forces on a particle. Drag forces on a particle have been expressed by R. P. Musselman (Ref. 5) as:

$$F_d = C_p \frac{V^2}{2} A$$

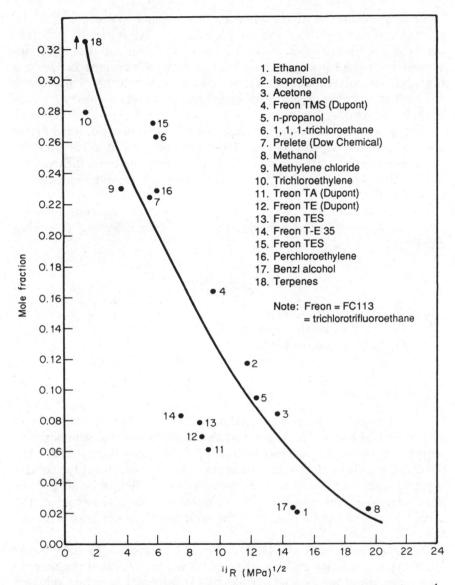

Figure 11.12 The mole fraction solubility of rosin (abietic acid) in various solvents.[4]

where F_d = drag force on particle
C = coefficient of drag
p = density of solvent
V = velocity of solvent
A = projected frontal area of particle

Upon analysis of the above equation, the density and velocity of the solvent are the only two controllable factors. High-density solvents can be chosen for proportionally higher drag force, and drag force increases with the square of the solvent velocity. Therefore, choosing a high-density solvent and spraying it at a high velocity will achieve the high drag forces on particles and contamination necessary for ease of their removal.

Once a solvent is chosen, we can consider its density a fixed factor leaving the velocity (pressure) of the applied solvent a controllable variable. Since cleaning under SMDs is usually the most difficult areas to clean, we must understand the effects of solvent velocities in these areas.

A model that describes the velocity of a solvent flowing, as a result of spraying, under a SMD can be explained by the following equation:[5]

$$V = \frac{(P_1 - P_2)y_2}{12ux}$$

where V = velocity of solvent
   P_1 = entry pressure of solvent
   P_2 = exit pressure of solvent
   y = clearance distance between SMD and substrate
   x = length of SMD
   u = solvent dynamic viscosity

Upon analysis of the above equation, which is sometimes referred to as the *law of two walls*, the solvent's viscosity and the component's length are increasingly proportional to the spray velocity under the SMD. This tells us that a low-viscosity solvent is essential for obtaining high-solvent velocities. Furthermore, as the length or size of the SMD increases, the solvent velocity decreases and, as expected, the solvent's velocity increases with the square of the distance between the SMD and substrate.

The equations that have been given which express the drag forces and solvent's velocity underneath a SMD are applicable when there is an open channel for solvent to flush under the SMD from one side and out the other. This condition, however, does not always happen, depending on the component-to-substrate clearance. Therefore, let us review what actually happens when flux residues are dispersed on a SMA and then solvent spray–cleaned.

In regard to leadless components such as chip resistors and capacitors, an open channel usually does not exist after the soldering process. Due to the small clearance, from 0.001 to 0.003 in. underneath leadless components, displaced flux will fill up the gap as shown in

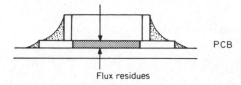

PCB

Flux residues

Figure 11.13 Displaced flux residues after soldering, filling the gap underneath leadless SMDs.

Fig. 11.13. Upon cleaning and utilizing solvent spray, the flux residues are removed in a progression of stages when components with such clearance are cleaned through an in-line system. Figure 11.14*a* through *e* illustrates what normally happens to the removal of flux residues from around and underneath leadless chip components. Primarily, the flux residues surrounding the SMDs body are removed first during the initial cleaning stages. As the flux residues soften due to the solvent's increase in temperature, solvency, etc., and are continually drained away, the mechanical effects of spraying start to remove the flux residues underneath the SMD. Once a open channel is produced, the solvent spray starts to flush the entire region underneath until the flux residues are removed.

When spray cleaning underneath SOICs and PLCCs, it is usually expected that an open channel for continuous solvent flushing occurs from the start. This is due to the higher component-to-substrate clearances offered by these SMDs. Actually, flux residue removal underneath these SMDs (44 leads or less) can be easier than the removal of flux residues from any type of leadless SMD.

11.4 Solvents

Solvents currently used in solvent cleaners for vapor-spray and spray-immersion decontamination of SMAs are usually organically based with a main chemical skeleton composed of carbon. These solvents can be classified under four groups: hydrophobic, hydrophilic, azeotropes of hydrophobic blends, and azeotropes of hydrophobic-hydrophilic blends.[1]

Hydrophobic solvents are not miscible with water and consequently exert little, if any, dissolving action with ionic contaminants. Due to this nonpolar characteristic, most hydrophobic solvents dissolve nonpolar or nonionic contaminants such as rosin, oils, and greases. Figure 11.15 provides a list of common hydrophobic solvents.

Hydrophilic solvents are miscible with water and can exert a dissolving action with nonpolar and polar contaminants. Figure 11.6 provides a list of common hydrophilic solvents; as shown, most of them

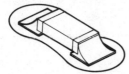

(A) Flux displaced around and underneath the SMD after soldering.

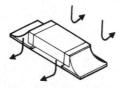

(B) Initial solvent spray cleaning stages removes the flux residues around the SMD.

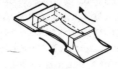

(C) The solvent's chemical and mechanical affects start to remove flux from underneath the edges of the SMD.

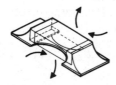

(D) The solvents spray progressively removes additional flux residues from underneath the SMD.

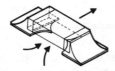

(E) Finally the solvent spray carves an open channel to permit the flushing of solvent from one side to the other side of the SMD.

Figure 11.14   The progressive steps of flux residue removal from a leadless SMD upon solvent spraying. (*Courtesy Electrovert Ltd.*)

Figure 11.15   **Common Hydrophobic Solvents**

Methylene chloride (dichloromethane)
Trichloroethylene
Perchloroethylene
1,1,1-Trichloroethane
Trichlorotrifluoroethane (1,1,2-trichloro-1,2,2-trifluoroethane)
Tetrachlorodifluoroethane
Trichlorofluoromethane

Figure 11.16   **Common Hydrophilic Solvents**

Methanol
Ethanol
n-Propanol
Acetone
Isopropanol
Methyl ethyl ketone
Methyl butanol
sec-Butyl alcohol

are alcohols. Usually hydrophobic solvents have a greater dissolving action for nonpolar contaminants, and hydrophilic solvents have a greater dissolving action for polar contaminants. To overcome these degrees of dissolving action for polar and nonpolar contaminants and to maximize the dissolving action for both types of contaminants, hydrophobic and hydrophilic solvents are mixed together to form azeotropic blends or solvent blends.

An azeotropic solvent mixture is composed of two or more different solvents which behave as a single solvent; i.e., the vapor produced by its partial evaporation has the same composition as each solvent that makes up the azeotropic solvent mixture. The azeotropic solvent mixture, which usually consists of one type of hydrophobic solvent and one or more types of hydrophilic solvent, has a constant boiling temperature. However, when the azeotropic solvent mixture becomes contaminated with flux residues, it behaves more like a solvent blend. Solvent blends can be treated like stable azeotropic mixtures; however, a periodic analysis should be maintained to recognize potential hazards such as flammability and inconsistent cleaning results.

The major ingredient of the most common solvent brands is either trichlorotrifluoroethane, tetrachlorotrifluoroethane, trichlorofluoromethane, or 1,1,1-trichloroethane. One of these hydrophobic solvents is then mixed with a smaller percentage of one or more hydrophilic solvents, usually alcohols, to the point where the mixture behaves as a nonflammable azeotropic mixture or solvent blend. The typical solvent brands sold commercially can be divided into two groups—chlorocarbon and fluorocarbon solvents. Fluorocarbon is the more common name used to describe chlorofluorocarbons. Reviewing these two groups of solvents in the following sections will allow us to compare their differences regarding to their advantages and disadvantages.

11.4.1   Chlorocarbon solvents

The major chlorocarbon solvents are 1,1,1-trichloroethane, perchloroethylene, trichloroethylene, and methylene chloride. These solvents and their azeotropic mixtures and blends with alcohols are generally characterized by higher solvency, higher boiling points, and lower cost than fluorocarbon solvents.

Early solvent cleaning processes used trichloroethylene-based solvents in vapor-spray degreasing systems. Due to toxicity, trichloroethylene has typically been replaced by safer and lower boiling 1,1,1-trichloroethane and its blends.

Perchloroethylene has the highest boiling point (121°C) of these four major chlorocarbon solvents. Due to this physical characteristic, it is normally only used at room temperature for cold cleaning flux residues. It is also commonly used as a cold cleaner in wave soldering finger cleaners, however, it is not recommended due to its toxicity.

Methylene chloride has the lowest boiling point (39.8°C) of the chlorocarbon solvents and has a very high solubility for organic contaminants such as rosin, oils, and greases. However, due to its toxicity and ability to easily react with a wide range of polymers, it is not recommended for use in high concentrations. Currently there are two azeotropic solvents incorporating methylene chloride—Genesolv DTA and Freon TMC, which are manufactured by Allied Chemical and Dupont, respectively. In any case, azeotropic mixtures or solvent blends incorporating methylene chloride should be expected to be extremely effective in the vapor-spray decontamination of flux residues. A precaution is strongly recommended: every type of component, PCB, and solder mask that may be cleaned with this chlorocarbon should be checked for compatibility before production or prototype use.

Currently, with no exception, the major chlorocarbon solvent used in azeotropic mixtures and blends is 1,1,1-trichloroethane. This is the safest of the four major types of chlorocarbon solvents, and when combined with polar solvents such as alcohols, provides excellent flux residue decontamination. For these reasons, the following sections deal exclusively with this chlorocarbon solvent. Figure 11.17 illustrates the physical characteristics of all four major types of chlorocarbon solvents.

Figure 11.17   Physical Characteristics of the Most Common Chlorocarbon Solvents

	1,1,1-trichloro-ethane (methyl-chloro-form)	Perchloro-ethylene	Trichloro-ethylene	Methylene chloride (dichloro-methane)
Chemical formula	CCl_3CH_3	$CCl_2{=}CCl_2$	$CCl_2{=}CHCl$	CH_2Cl_2
Molecular weight	133.4	165.9	131.4	84.9
Boiling point	166.4°F (74.67°C)	250°F (121°C)	188°F (86.67°C)	103.64°F (39.8°C)
Specific gravity, 25/25°C, lb/gal	1.312	1.620	1.460	1.316
Density liquid, 25°C, lb/gal	11.03	13.50	12.10	10.94
Sat. vapor density at bp (air = 1.00 lb/ft^3)	4.50	5.83	4.54	2.93
Vapor pressure, 20°C, mmHg	175.0	14.0	58.6	350.0
Viscosity, 25°C, cP	—	—	—	0.329
Surface tension, 20°C, dyn/cm^2	25.90	32.30	29.20	28.12
Latent heat of vaporization, Btu/lb	102.0	90.0	103.0	141.7
Specific heat, Btu/lb, °F, liquid	0.26	0.21	0.23	0.28
Evaporation rate, carbon tetrachloride = 100	139	27	69	147
Solubility, 25°C, wt %: Water in solvent	—	0.01	0.02	0.20
Solvent in water	< 0.10	0.02	0.11	2.00
Kauri-butanol value, cm^3	124	92	130	136
Flash point	—	—	—	—
Threshold limit value, ACGIH	350	50	50	100

11.4.1.1   Chlorocarbon solvent: 1,1,1-trichloroethane.   Currently, Prelete (London Chemical) and Alpha 565 (Alpha Metals) are the two commercially available chlorocarbon solvents containing 1,1,1-trichloroethane as a major ingredient. The physical characteristics of these two chlorocarbon brands are given in Fig. 11.18.

Figure 11.18   Physical Characteristics of the Most Common Chlorocarbon Base Solvents Containing 1,1,1-Trichloroethane

Solvent brand name	Prelete	Alpha 565
Manufacturer	London Chemical	Alpha Metals
Appearance	Colorless liquid	Colorless liquid
Boiling point, 760 mmHg	164°F (73.3°C)	165°F (73.9°C)
Lower flammable limit, 77°F	7.0%	3.0%
Upper flammable limit, 77°F	15.5%	15.0%
Flash point, T.C.C.	None	None
Surface tension, dyn/cm	25.2	—
Viscosity, 25°C, cSt	0.618	—
Specific gravity, 77°F	1.286	1.275
Vapor density (air = 1)	4.16	4.6
Ingredients:		
1,1,1-Trichloroethane	92.9%	90%
1,2-Butylene oxide	< 7.1%	—
Diethylene ether	< 7.1%	< 5%
Methyl butanol	< 7.1%	—
Nitromethane	< 7.1%	—
sec-Butyl alcohol	< 7.1%	—
n-Propanol	—	< 5%

To improve their ability to remove polar contaminants, alcohols are added which give both Prelete and Alpha 565 solvents very good and similar performances in flux residue decontamination. Their boiling points, which are significantly higher than most fluorocarbon solvents, greatly decrease the time to remove flux residues. When batch cleaning is used or when fast conveyor speeds (> 5 ft/min) are necessary, this can be a very effective advantage over most fluorocarbon solvents. These higher temperatures may create a minor assembly handling concern just after cleaning; however, if necessary, putting on a good pair of thin antistatic gloves soon after cleaning would assist those whose hands are very sensitive to heat.

The compatibility of 1,1,1-trichloroethane with polymers (plastics, elastomers, and inks used in marking) depends on a number of factors including the chemical properties of the polymer in question, the polymer formation parameters, the degree of cross-linking (curing or vulcanizing conditions), and the time of exposure. Prelete and Alpha 565 solvents have been shown to be compatible with most polymers used in the manufacture of PCBs and components including all plastics that conform to MIL-STD-202. Neither of these solvent brands is compatible with styrene-based polymers and polycarbonates. With aluminum electrolytic capacitors, chlorocarbon- and fluorocarbon-based solvents should only be used if the capacitors are sealed at the lead-body interface with an epoxy barrier coating. Variable resistors or tuning-type components, which may have unsealed lubricating oils or greases for decreasing material wear, are not compatible with any solvent-based cleaner. Finally, relays and their performance changes should be thoroughly tested when 1,1,1-trichloroethane–based solvents are used.

The other concern with 1,1,1-trichloroethane–based solvents is the toxicity. This chlorocarbon solvent is among the most widely studied and commercially used chemicals in today's manufacturing environments. On the basis of all the data now available, it is reasonable to assume that azeotropic mixtures or solvent blends containing 1,1,1-trichloroethane present a negligible risk to operators whose exposure does not exceed currently acceptable exposure guidelines provided in the manufacturers' Material Safety Data Sheets. However, chronic overexposure to 1,1,1-trichloroethane has caused toxic liver effects in experimental animals. Diethylene ether, a stabilizer used in Alpha 565 solvent, has been identified as an animal carcinogen by NTP and IARC, but not by OSHA. Adding 1,1,1-trichloroethane to diethylene ether has been tested in laboratory animals and not found to cause cancer.

11.4.1.2 Fluorocarbon solvents. As mentioned, the most common fluorocarbon solvents used as the major ingredients in azeotropic mixtures or solvent blends are trichlorotrifluoroethane (1,1,2-trichloro- 1,2,2-trifluoroethane), tetrachlorodifluoroethane (1,1,2,2-tetrachloro- 1,2-difluoroethane), and trichlorofluoromethane. These are commonly referred to as Freon-based solvents with so-called refrigerant numbers FC-113, FC-112, and FC-11, respectively. Of these three, only FC-113 and FC-112 are used as the major ingredients in azeotropic mixtures or solvent blends since FC-11's low boiling point (23.82°C) and high vapor pressure make it too volatile for ordinary use. FC-11 is sometimes used as a very low toxicity cold cleaning solvent. Figure 11.19 lists the physical characteristics of these three major types of fluorocarbon solvents.

The FC-113 fluorocarbon solvent, sometimes referred to as Freon TF, is the most commonly used fluorocarbon solvent, whether in pure form or as a azeotropic mixture or solvent blend. In pure form, FC-113 has a low solubility for rosin or nonpolar contaminants; therefore, it requires the addition of a highly polar solvent such as alcohol. Due to FC-113's mild solvency power, it has very little effect, if any, on polymers when comparing it to the chlorocarbon-based solvent, 1,1,1-trichloroethane.

The FC-112 fluorocarbon solvent has a high boiling point of 92.8°C, whereas FC-113 has a boiling point of 47.6°C. This physical characteristic, plus the fact that it has excellent solvency of nonpolar contaminants, has made the FC-112 azeotropic mixture an outstanding vapor-

Figure 11.19   Physical Characteristics of the Most Common Fluorocarbon Solvents

Characteristic	Fluorocarbon solvent		
	FC-113	FC-11	FC-112
Chemical formula	CCl₂F—CF₂Cl	CFCl₃	CCl₂—CCl₂F
Molecular weight	187.4	137.4	203.9
Boiling point, 760 mmHg	117.6°F (47.6°C)	74.7°F (23.7°C)	199°F (92.8°C)
Density of liquid, 25°C:			
lb/gal	13.06	12.39 (21.1°C)	—
g/cm³	1.565	1.485 (21.1°C)	1.645
Saturated vapor at bp, air = 1.0	6.20	—	7.47
g/L	7.399	—	—
lb/ft³	0.4619	0.3806	—
Vapor pressure, 21.1°C, mmHg	472	1149.7	46 (20°C)
Viscosity, 21.1°C, cP, liquid	0.694	0.438	1.06 (20°C)
Surface tension, 20°C, dyn/cm	18.8	19.0	23.9
Latent heat of vaporization at bp:			
Btu/lb	63.1	78.3	—
cal/g	35.06	43.5	41.0
Specific heat at 21.1°C:			
Btu/(lb)(°F), liquid	0.21	0.21	0.22 (20°C)
Saturated vapor	0.15	0.13	—
Evaporation rate (carbon tetrachloride = 100)	280	225	64
Solubility, wt %, 25°C:			
Water in solvent	0.011	0.009	0.009 (20°C)
Solvent in water	9.017	0.011	0.012 (20°C)
Kauri-butanol value, cm³	34	60	70
Thermal conductivity, liquid, Btu/(h)(ft³)(°F/ft)	0.0449	0.0630	0.0400
Dielectric constant, 60 Hz, liquid, 30°C	2.41	2.20	—
Flash point	None	None	None
Threshold limit value, ACGIH	1000	1000	500

spray decontamination solvent. Like that of 1,1,1-trichloroethane, its higher boiling point allows effective use of FC-112 azeotropes in high volume production where the rate of product throughput must be as fast as possible.

There are many different types of azeotropic mixtures and solvent blends commercially available under different brand names which have FC-113 as the major ingredient. Currently, there are two azeotropic mixtures containing FC-112, both of which are both manufactured by Alpha Metals. Figure 11.20 provides the physical characteristics of azeotropic mixtures and solvent blends containing FC-113 or FC-112 as the major ingredient. Figure 11.21 provides a list of the ingredients found in some of the more common brands of fluorocarbon solvents.

11.4.1.3 Terpene solvents.

As the result of a recent international agreement among 37 nations called the Montreal Protocol, the production of fluorocarbon (or chlorofluorocarbon) solvents is being cut back to reduce the effects these solvents have on stratospheric ozone depletion. The Montreal Protocol, which has been accepted by solvent manufacturers such as Dupont, scheduled cutbacks in solvent production to revert to 1986 levels and by 1999 to cut back to half the 1986 levels.

Of course, to realistically accomplish the real goal of reducing the damage to the earth's stratospheric ozone layer, a replacement solvent with little or no reaction with ozone must be found and made acceptable for use. As a result of recent research, a class of naturally derived solvents chemically similar to abietic acid called *terpenes* have emerged as "possible" alternatives to fluorocarbon solvents. Terpenes are widely found in nature and occur in nearly all living plants. The two most abundant sources of terpenes are turpentine and other essential oils. Terpenes are generally regarded as derivatives of isoprene (2-methyl-1,3-butadiene); are readily biodegradable; and are free of the ozone-damaging halides, noncorrosive, and essentially nontoxic.

A commercially available terpene solvent targeted to clean electronic assemblies, called Bioact EC-7, has been developed by Petroferm, Inc. (Fernandina Beach, Florida). This solvent contains the terpene lionene as the major ingredient as well as a surfactant to make this terpene solvent rinsable in water as a final, required step in the cleaning process.

Terpenes are classified as combustible (flammable) substances since most have a flash point between 100 to 200°F (Tag Close Cup). Since EC-7's flash point is 117°F, it must be used as a cold cleaning solvent, and dumping used solvent into city drainage would be prohibited. Recycling used EC-7 will also present some difficulties since typical dis-

Figure 11.20 Physical Characteristics of Several Common Azeotropic Mixtures and Solvent Blends Containing Fluorocarbon Solvents

Characteristic	Alpha 1001	Alpha 1003	Freon TMC	Freon TMS	Genesolv DTA	Genesolv DMS	Genesolv DES	Genesolv DFX
Boiling point, 760 mmHg:								
°F	168	180	97.2	103.5	93.5	107.6	112.0	
°C	76	82	36.2	39.7	34.2	42.0	44.4	
Latent heat of vaporization at bp, Btu/lb	128	100	104.0	90.7	—	—	78.0	
Density, g/cm^3 (a = 68°F, b = 77°F)			1.420[b]	1.477[b]	1.404[a]	1.462[a]	1.486[a]	
Viscosity, cP (a = 70°F, b = 77°F)			0.461[b]	—	0.538[b]		0.70[a]	
Surface tension, dyn/cm (a = 75°F, b = 77°F, c = 70°F, d = 68°F)			21.4[b]	17.4[b]	22.1[c]	19.0[d]	19.4[a]	
Kauri-butanol number	71	71	86	45	148	48	43	

Figure 11.21   Some Common Fluorocarbon Azeotropic Mixtures and Solvent
Blends with Their Types of Ingredients

Azeotropic Mixtures, %			
Alpha 1001[*]:		Alpha 1003[*]:	
Tetrachlorodifluoroethane	—	Tetrachlorodifluorethane	—
Isopropanol	28	n-Propanol	15
Stabilizer	—	Stabilizer	—
Freon TMC[†]:		Genesolv DTA[‡]:	
Trichlorohifluoroethane	50.5	Trichlorotrifluoroethane	55.0
Methylene chloride	49.5	Methylene chloride	41.7
		Methanol	3.2
Solvent Blends, %			
Freon TMS[†]:		Genesolv DMS[‡]:	
Trichlorotrifluoroethane	94.05	Trichlorotrifluoroethane	92.0
Methanol	5.70	Methanol	4.0
Nitromethane	0.025	Ethanol	2.0
		Isopropanol	1.0
		Nitromethane	1.0
Genesolv DES[‡]:		Genesolv DFX[‡]:	
Trichlorotrifluoroethane	93.5	Trichlorotrifluorethane	
Ethanol	3.5	Methanol	
Isopropanol	2.0	Stabilizer	
Nitromethane	1.0		

*Alpha Metals.
†Dupont.
‡Allied Chemical.

tillation procedures for noncombustible solvents require boiling the solvent, something extremely dangerous with a combustible solvent.

Although terpene solvents show excellent dissolving capabilities for rosin, its combustibility remains the major factor limiting its use. Since it must be used as a cold solvent, its rate of dissolving action on solid flux residues could be limited for high-volume production, and the fact that a separate final water rinse is mandatory for removing surfactants could further limit its use for cleaning SMAs.

The author feels that efforts to find suitable replacement solvents that have little or no effect on the stratospheric ozone layer are quite warranted and perhaps mandatory for our environmental safety. However, using combustible solvents in the typical manufacturing environment should be undertaken with extreme caution since flammable aerosols can also be explosive.

11.4.1.4   Stability of solvents.   Solvent stability is essential for dependable performance. If solvents cause a chemical reaction or decompose during use, damage to materials in contact with the solvent become widespread and expensive, and solvent recovery and reuse become impractical or impossible. Although some of the solvents discussed in this chapter have been advertised by manufacturers to be stable, to various degrees, they all can react and cause corrosion with active metals such as aluminum, beryllium, magnesium, and zinc during extended exposure in the presence of alcohols. The chances of such exposure occurring in the manufacturing process have increased due to the multiple cleaning process sometimes necessary with certain surface mount designs. Also, using batch solvent systems, which may have no automated controls for assembly dwell times, and long, conveyorized, solvent immersion systems can easily contribute to extended solvent exposure times and corrosion of active metals.

Fortunately, this problem can be overcome by incorporating the correct type and amount of inhibitors into the solvent. These inhibitors distribute themselves in both the vapor phase and boiling sump chambers of the cleaning system, according to their boiling points. The inhibitors prevent the corrosion of the cleaning systems and metal parts being processed. The three types of inhibitors that can be used in a solvent are:

1. An *acid acceptor,* which reacts with and chemically neutralizes trace amounts of hydrogen chloride formed during solvent use.

2. A *metal stabilizer,* which deactivates the metal surface and complexes any metal salts that might form.

3. An antioxidant, which reduces the solvent's potential to form oxidation products.

In most cases, metal stabilizers are added to fluorocarbon solvents, whereas both an acid acceptor (such as 1,2-butylene oxide) and stabilizers (such as diethylene ether) are sometimes added to chlorocarbon solvents. Several types of inhibitors that are added to solvents to help deactivate metal surfaces are listed in Fig. 11.22.

The effects of metal-solvent-inhibitor interaction can be explained with the condition that occurs when 1,1,1-trichloroethane comes in contact with aluminum. As shown in the following reaction, 1,1,1-trichloroethane reacts with aluminum to produce a saturated dimer and aluminum chloride:

$$2Al + 6CH_3CCl_3 \rightarrow 3CH_3CCl_2CCl_2CH_3 + 2AlCl_3$$

When correctly present in solution with the 1,1,1-trichloroethane, the inhibitors will form an insoluble complex with the aluminum chloride or other type of metal chloride which is formed at the initial microscopic reaction sites. This insoluble complex then acts as an effective

Figure 11.22   Inhibitors Commonly Used in Solvents to Prevent Corrosion of Metals

| Inhibitor | Solvent | |
	Fluorocarbon	Chlorocarbon
1,2-butylene oxide		X
Nitromethane	X	
1,4-Dioxane	X	
sec-butanol	X	
1,3-Dioxolane		
tert-butanol		
Methyl butynol	X	
Isopropylnitrate		
Acetonitrile		
Diethylene ether	X	X

physical barrier separating the two reactive components, solvent and metal.

11.4.2   Solvent performances

By now most of us have read advertisements and technical papers that claim how one manufacturer's solvent will outperform another. In all cases, a given solvent manufacturer will never find someone else's solvent to be a better one; actually, what has happened is an increase in confusion among users and potential users of these solvents. In fact, some solvent users blindly support a particular solvent type rather than what is best overall for manufacturing and, as a result, report conditions that seem quite distorted.[4]

As discussed in more detail in Sec. 11.4.6, a solvent's performance can be measured by testing the amount of ionic contaminants left on the assembly after it has been cleaned with the respective solvent. A test such as this can be performed using one of the several solvent extract conductivity-resistivity machines available on the market. To compare some of the more commonly used solvents, solvent extract conductivity tests, using an Ionograph 500, were performed on standardized SMAs with various manufacturers' solvents and fluxes as shown in Fig. 11.23.

The test assemblies were composed of several leadless chip carriers, SOICs, PLCCs, gull-winged PLCCs, and chip resistors. The I/O count on the active components ranged from 14 to 64. The substrate was 6 × 7 in., 0.062-in.-thick FR-4. The cleaning process consisted of 10 assemblies for each type of flux used in the soldering process and for each solvent tested. Each assembly underwent a dwell time of 5 min in the saturated vapor zone of a batch cleaner.

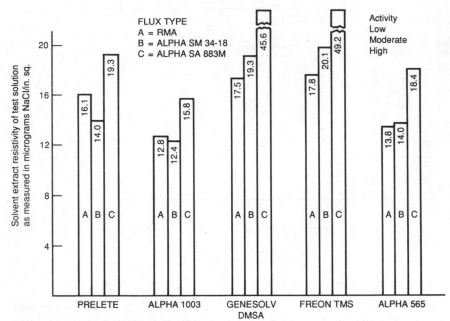

Figure 11.23 Levels of cleanliness achieved by various solvent brands when cleaning SMAs soldered with different fluxes.

As shown in Fig. 11.23, the tetrachlorodifluoroethane base solvent (Alpha 1003) outperformed all other solvents. The 1,1,1-trichloroethane base solvents (Alpha 565 and Prelete) outperformed the solvents containing trichlorotrifluoroethane-based solvent (Freon TMS and Genesolv DMSA).

Based on this information, one may deduce that the best solvent is Alpha 1003. This may not be entirely true since this solvent has a high boiling point of 180°F, which may have an effect on some of the polymers used on the assembly and also one's cleaner may not have immersion heaters capable of supplying enough power to boil the solvent. Furthermore, Alpha 1003 is approximately three times the cost of the other types of solvents. This may not be a factor, since the vapor losses of the Alpha 1003 solvent are 63 percent and 90 percent less than solvents containing 1,1,1-trichloroethane and trichlorotrifluoroethane, respectively. However, its density is also higher, which may produce more solvent dragout.

When comparing the 1,1,1-trichloroethane–containing solvents with those containing trichlorotrifluoroethane it is obvious that solvents which contain 1,1,1-trichloroethane are significantly better at cleaning assemblies soldered with a highly activated synthetic flux. Also, due to 1,1,1-trichloroethane's higher boiling point temperature than trichlortrifluoroethane, Alpha 565 and Prelete can clean a lot faster than either Freon TMS and Genesolv DMSA. This can be a significant factor in batch cleaning large quantities of assemblies and in high-speed, high-volume manufacturing. Solvents containing trichlorotrifluoroethane also have some advantages such as lower toxicity and their compatibility with a wider variety of polymers, which may be the ruling factors for some people in choosing the "best" solvent.

In all cases, when choosing the solvent that best suits the user's needs, the following questions should be answered:

1. Does the solvent provide an effective amount of decontamination under the SMDs being used?

2. Can the solvent perform effective decontamination in the time given to perform the cleaning operation?

3. Does the solvent affect any component or material on the assembly?

4. Does the solvent have an acceptable toxicity level? (All solvents are toxic.)

5. Are solvent losses caused by environmental losses and dragout economically acceptable?

6. Is the solvent compatible with the cleaning system? When PVC plumbing is used, consult the solvent manufacturer concerning compatibility.

7. Does the solvent provide, over time, acceptable levels of stability? Based on the author's experience, none of the solvent brands that exist on the market today behave as true azeotropic mixtures once the solvent in the boiling sumps becomes contaminated with flux residues. Eventually, all solvent brands behave as solvent blends during use and exhibit variations related to their percentage of changes in ingredients. It is best to periodically remove solvent samples from each sump so that gas chromatograph testing can verify the change in the solvent's composition.

11.4.3 Storage of solvents

Solvents are typically packaged in 5-gal pails, 55-gal drums, or bulk storage tanks. When a small batch cleaning system is used, 5-gal pails are usually sufficient to handle this kind of solvent usage. If the batch system is heavily used or if an in-line cleaning system is used, 55-gal drums are preferred. When drum usage increases to three drums or

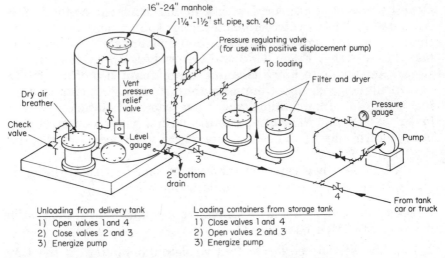

Figure 11.24 Schematic diagram of a bulk storage tank system for solvents.

more per month, bulk storage will be simpler and more economical. Having 55-gal drums hauled off can cost $50 to $150 per drum, depending on the location.

A bulk storage tank, which is sometimes offered free of charge from solvent suppliers (of course, you must buy their solvent), can be located indoors or outdoors, depending on the local codes on solvent usage and handling. The tanks usually are available in sizes from 250 gal and up. Solvent deliveries from transport trucks usually require 200 gal minimum.

Figure 11.24 illustrates the schematic diagram of a typical tank storage system. The in-line filter and dryer are optional and are seldom seen on less expensive small tanks. The pressure and vacuum relief valve, safety seal, and vent drier are recommended for all tank installations. Note that the filling point is on the bottom of the tank. This, or a submerged fill pipe from the top of the tank, is required by EPA legislation in many locations.

11.4.4 Solvent cleaning systems

11.4.4.1 Batch solvent cleaners. Batch solvent cleaners or degreasers have been used to successfully clean SMAs as an effective, less expensive alternative to in-line (conveyorized) solvent cleaners. In the past few years, advancements in solvent cleaning techniques and instrumentation have enhanced the basic process. Furthermore, innovations and refinements in process monitoring and documentation can significantly affect the ability to achieve assembly cleaning levels easily reached by in-line solvent cleaners.

Many different types of batch cleaners have been designed. However, the four basic designs are the perimeter batch system, the offset batch system, two-tank batch systems, and three-tank batch systems. These systems are illustrated in Fig. 11.25.

Whether batch or in-line, all solvent cleaning systems employ a built-in solvent distillation cycle. This distillation cycle is described in the following steps:

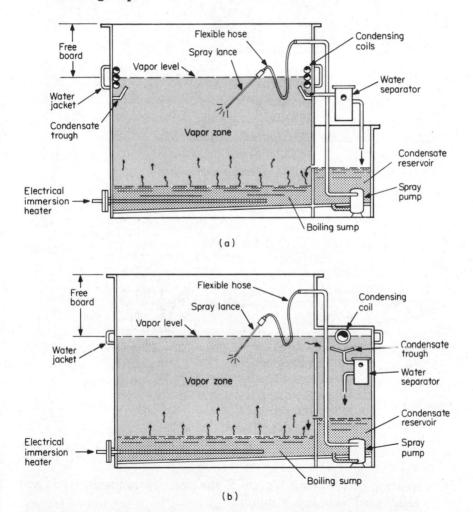

Figure 11.25 The four major types of solvent batch cleaners. (a) Perimeter batch system; (b) offset batch system; (c) two-tank batch system; and (d) three-tank batch system. (*Courtesy Diamond Shamrock Chemical Company.*)

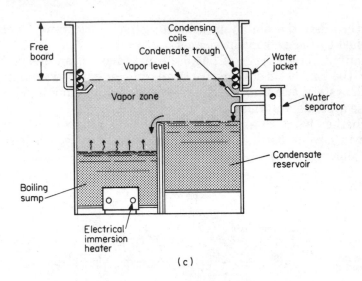

(c)

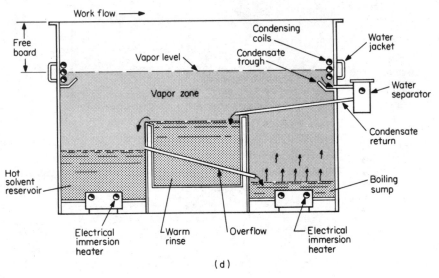

Figure 11.25 *(Continued)*

1. Vapor is generated from the boiling chamber(s) by electric immersion heaters.

2. Vaporized solvent thus travels to the primary condenser coils and is condensed back into a liquid.

3. Liquid distillate then travels by piping to the solvent drying unit where it flows through a water removal unit for moisture removal. Water removal units should always be incorporated; however, on some systems they are optional.

4. Dried solvent then flows by piping to the distillate reservoir where it is used for distillate spray via pumping.

5. An overflow pipe or retaining wall then directs the solvent back to the boiling tank(s) for reboiling.

From an efficiency standpoint, there is still another type of batch system, which, instead of using electrical conversion heaters to boil the solvent and chill the water to condense the solvents, utilizes a patented variable heat refrigeration system to do the same. First developed by Corpane Industries, the variable heat refrigeration system provides a significant reduction in utility costs when compared to conventional systems.

As further explained in Sec. 11.4.4.3, the system employing variable heat refrigeration uses the hot side of the refrigeration unit to boil the solvent and the cold side to condense the solvent. This in itself limits any extra energy loss, which is not as easy to control with the conventional systems.

Operating a batch system should be tightly controlled and well enforced by supervision. A poorly documented and controlled process will undoubtedly yield very poor cleaning results.

Prior to each use or at the beginning of each shift, a batch cleaner should be checked and set up as follows:

1. The boiling sump that collects the flux residues should have a sufficient amount of solvent to promote uniform and rapid boiling so that a saturated vapor zone is easily maintained.

2. If a workstand is placed in the boiling sump to support the cleaning load, always keep the contaminated solvent a safe level below the horizontal shelf of the workstand. This will keep the bottom of the basket used to lower and raise the cleaning load from carrying over contaminated solvent into the other solvent tank(s).

3. Fill and maintain the solvent level in the other tank(s) so that the solvent is always cascading over into the boiling sump tank.

4. When the machine is turned on, allow sufficient time (usually a minimum of 15 min) for a saturated vapor zone to occur and check to make sure the condensing coils are as cold as prescribed by the manufacturer's manual.

5. Depending on the amount of use, it is essential to replace the solvent in the boiling sump with fresh solvent periodically. Consider scheduling this type of maintenance at least weekly for large volumes of work.

Once the proper machine setup conditions have been accomplished, maintaining good process controls will produce good cleaning results.

The best process controls for operator use are the simplest ones. The following list of process and operator controls should be the minimum to incorporate and enforce religiously:

1. The operator should always wear eye protection, preferably goggles. A small drop of solvent in the eye can result in a severe medical problem.

2. Assemblies should be loaded vertically on trays and slowly lowered into the vapor zone over the boiling sump (usually the tank farthest over to the left side of the machine). Immersion of the assemblies in the boiling sump tank is strongly not recommended. The cleaning systems currently available do not provide sufficient solvent recovery rates to maintain low levels of dissolved flux residues, at least in the first tank where most of the flux residues end up.

3. Trays which hold the assemblies are normally lowered in baskets and are allowed to rest on the workstand sitting above the boiling solvent. The dwell time in this position should be a minimum of 60 s or until the solvent stops condensing on the assemblies.

4. After the vapor dwell, a good close spraying of each board should be performed with a hand-held spray wand activated by a foot pedal.

5. The assemblies should sit for another period of time (usually 60 to 90 s) in the solvent vapors. Usually during spraying the solvent vapor blanket collapses; therefore, allow extra time for the saturated vapor zone to reform.

6. Once the cycle is completed, the operator should slowly pull the basket of assemblies vertically out of the cleaner.

7. As often as possible and after machine shutdown, the lid should be placed over the machine to prevent extra solvent losses to the atmosphere.

11.4.4.2 In-line solvent cleaners.   In-line solvent cleaners are used for minimizing operator variations and for low- to high-volume production cleaning of assemblies. A typical in-line cleaner consists of a long vapor chamber tank subdivided into smaller tanks designed to accommodate solvent cascading, solvent boiling, reservoirs for solvent spray, and sometimes for immersing assemblies in boiling solvent.[5] The assemblies normally travel horizontally through the vapor chamber on a continuous conveyor at various speeds depending on the type of SMA. Of course, the same solvent distillation and condensation cycle occurs within in-line cleaners as explained in Sec. 11.4.4.1 for batch cleaners.

The key to clean SMAs with in-line cleaners is to select the preferred solvent and, most importantly, since the solvent type can always be changed, to select the best cleaning cycle. The cleaning cycle is usually dependent on how the in-line cleaner is designed, and, of course, there are many different types of machine designs from various manufacturers.

Currently, there are three major cleaner designs which provide various types of cleaning cycles. The in-line cleaners which provide these cleaning cycles can be described as follows:

1. *Vapor-spray-vapor cycle:* This is the most commonly used cycle for in-line solvent cleaning. Figure 11.26 illustrates the cleaner design which provides this type of cleaning cycle. Preferably, the assembly first enters the vapor zone and then is immediately sprayed on the top and bottom sides. No matter what cleaning cycle is used, it is always preferable to spray the assembly prior to any other sequence in the cycle except vapor exposure. Initial and final spraying on an incline is also preferred because this helps increase the velocity of the solvent spray under the SMDs. This type of vapor-spray-vapor cycle is greatly enhanced with high-pressure spraying, which is typically classified as spray pressures ranging from 60 to 200 psi. Spraying can be accomplished by using a combination of flat spray, narrow fan spray, or wide fan spray nozzles. Figure 11.27 illustrates these three different types of spray nozzles. The final stages of this cycle should include horizontal solvent spraying, incline spraying, final vapor exposure, and solvent drainoff.

2. *Spray–boiling immersion–spray cycle:* In-line solvent cleaners designed to incorporate this kind of cycle have been designed primarily for difficult-to-clean SMAs. This cycle includes an initial incline

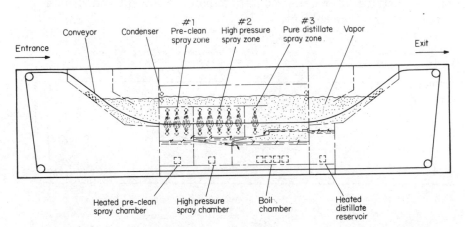

Figure 11.26 In-line solvent cleaner design for a typical vapor-spray-vapor cycle. (*Courtesy Detrex Chemical Industries, Inc.*)

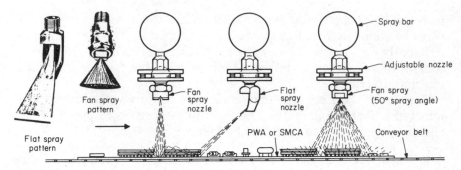

Figure 11.27 Flat spray, narrow fan spray, and wide fan spray nozzles. (*Courtesy Detrex Chemical Industries, Inc.*)

spraying, then assembly immersion in boiling solvent, then a final incline spray, and solvent drain.

3. *Spray–boiling immersion with spray–spray cycle:* This in-line cleaner has the same design as the spray–boiling immersion–spray in-line cleaner with the addition of solvent spray just above the boiling solvent. Some are also designed with spray nozzles submerged in the boiling solvent to create additional solvent turbulency. In this type of boiling immersion system, the solvent level can be lowered below the conveyor, which changes the cycle into a vapor-spray-vapor cycle. Figure 11.28 illustrates the cleaner design capable of providing these types of cleaning cycle combinations.

11.4.4.3 Variable heat refrigeration used in solvent cleaners. As previously mentioned for batch systems, the most efficient in-line solvent cleaners incorporate a variable heat refrigeration system which replaces electrical immersion heaters to boil the solvent and chilled water to condense the solvent.

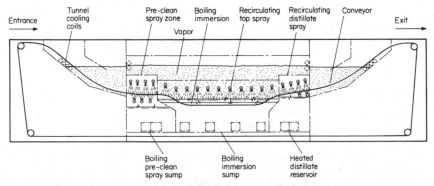

Figure 11.28 In-line solvent cleaner design for a typical spray–boiling immersion with spray–spray cycle. (*Courtesy Detrex Chemical Industries, Inc.*)

The normal heating and cooling cycle encountered in the conventional batch and in-line systems includes boiling and vaporization of the solvent in the sump(s) by means of electrical immersion heaters. Cooling the solvent vapor is accomplished by helical pipe coils located at the desired solvent vapor level. The solvent vapors condense back to liquid, after which the condensed liquid is returned to the boiling sump for reboiling and revaporization. Cooling conventional systems is accomplished by circulating water or refrigerant through the cooling coils. For high-volume production, a separate solvent recovery still is required for continuous distillation (purification) of solvent. The solvent recovery still, which is required for conventional systems, is used in conjunction with the solvent cleaning system for reducing contamination buildup in the solvent. Conventional batch systems usually have from one to three tanks of solvent from which vapor is generated. When spraying is required, a spray wand is used to spray cold solvent from the second or third tank on the work piece.

Variable heat recovery refrigeration systems operate on the same solvent cycle as described above, but without the use of a separate heater to boil or vaporize the solvent. A refrigerant is passed through a closed-loop plumbing system to boil and condense the solvent. In this case, electrical immersion heaters are not used for boiling the solvent for vaporization and circulating water in coils for vapor condensation.

A standard closed-loop variable heat recovery refrigeration solvent cleaning system operates on the same principle as standard refrigeration cooling systems. The heating (boiling of solvent) and cooling (condensation of solvent) cycles of the variable recovery refrigeration cleaning systems are shown in Fig. 11.29. In the condenser (2), refrig-

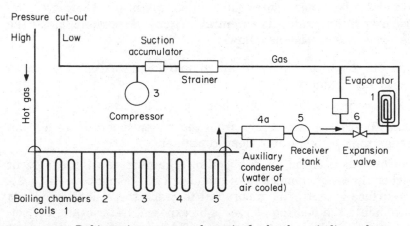

Figure 11.29 Refrigeration system schematic for batch or in-line solvent cleaners utilizing variable heat recovery technology.

erant that enters as liquid evaporates and absorbs heat, changing to a low-pressure, heat-laden gas. This part of the cycle is where the solvent condenses on the cooling coils.

The compressor (3) pumps the low-pressure, heat-laden gas (refrigerant) out of the evaporator and into the compressor cylinder. Here the gas is compressed by the action of the pistons and is delivered through the discharge tube as a gas under pressure to the helical coil(s) (4).

Here the compressed gas gives up its latent heat to the surrounding liquid solvent, causing it to boil. The gas turns to a liquid while continuing on to an auxiliary condenser (4a), sized to remove any remaining heat introduced to the gas by mechanical and electrical inefficiencies of the compressor.

When the gas loses heat, it reverts back to a liquid. This change of state is complete at the last pass of the auxiliary condenser. The liquid then passes from the auxiliary condenser to the receiver tank (5). The tank stores liquid for use by the evaporator (1).

Between the condenser and the evaporator is an expansion valve (6) which controls the flow of liquid according to the demand of the evaporator. From there, the liquid returns to the evaporator to repeat the cycle of heating and cooling.

11.4.5 Solvent cleaner maintenance

Solvent cleaner cleanliness is essential for operator safety, economy of operation, and efficiency of process. Factors determining the frequency of cleaning include the volume of work, type and amount of soil, and boiling temperature of solvent-flux mixtures.

An excessive accumulation of flux residue and other waste byproducts in the cleaner invites solvent decomposition. Therefore, routine cleaner maintenance is essential. Whenever possible, clean the cleaner from the outside as follows:

1. Turn off the heat supply to all compartments except the vapor-generating chamber.

2. Close solvent distillate return lines to the degreaser.

3. Route the condensate return from the water separator to a clean storage tank or a suitable clean container.

4. Continue distillation in the boiling chamber until the solvent reaches the recommended minimum level of 1½ in. above the heating surface or until the vapors fail to reach the condenser. At no time should the heating elements be exposed. Deterioration of solvent and destruction of the heater could result if this is allowed to happen.

5. Turn off the heat supply to the boiling chamber and allow the residue to cool before draining.

6. If residues must be drained prior to cooling, wear approved respiratory equipment and provide adequate ventilation.

7. Insoluble waste material should be removed with a hoe or similar tool. Particular attention should be given to the area under the heating coil.

8. Replace the plug in the drain line or close the drain valve. Open the solvent return line to the cleaner and adjust the liquid level in all compartments with distilled solvent, using additional fresh solvent as necessary.

9. The preferred options for disposal of solvents are to send the waste to a licensed reclaimer or to permitted incinerators, in compliance with local, state, and federal regulations, including Subtitle C of the Resource Conservation and Recovery Act. Dumping into sewers, on the ground, or into any body of water is strongly discouraged and is illegal.

Caution: *Do not store solvent-sludge mixtures in tightly closed drums. Occasionally, mixtures of waste and solvent can undergo reaction, generating gases and pressure that can burst standard steel drums.*

Comprehensive degreaser cleaning should be performed periodically. This necessitates the removal of all solvent from the degreaser and still. It also requires the services of competent maintenance persons who are thoroughly familiar with solvents and the hazards associated with the process. The procedure is as follows:

1. Transfer all solvent from the cleaner to storage tanks or suitable containers and turn off lines. It is preferable to distill all dirty solvent prior to cleaning. Observe minimum solvent levels as prescribed to prevent exposing the heating elements.

2. Turn off the heat and allow the unit to cool. Drain residual solvent and sludge completely from all tanks on compartments. Lock off electric conveyors.

3. Thoroughly aerate the unit by forced ventilation from a fan, compressed air, or blower, expelling solvent-laden air from the building via a suitable exhaust system.

4. Remove or protect indicating thermometers and controls to prevent damaging them.

5. Disconnect the heat supply and remove steam or gas coils or electric heaters.

6. Remove clean-out ports, taking care not to damage the gaskets.

7. Use a scraper or hoe to remove all dirt, sludge, and other insoluble waste from the bottom of each compartment. It may be necessary for maintenance personnel to enter some units. Such entry should be permitted only when the cleaner has been thoroughly ventilated and with maintenance personnel suitably protected.

8. When the cleaner has been thoroughly ventilated, a maintenance worker may begin to clean the condensate trough, condensing coils, and walls of each compartment. Care should be exercised to avoid damaging any corrosion-resistant finish on the interior of the unit. The degreaser should be checked for rusted areas and defects.

9. Thoroughly scrape the removed steam coils or electric elements to remove caked sludge that may interfere with heat transfer and disrupt the heat balance.

10. Clean the condensate line from the collecting trough to the water separator by blowing it out with compressed air or swabbing it with a wire brush.

11. For proper cleaning of the water separator, remove it if necessary.

12. Clean and check the various controls, indicators, and regulators before reassembly.

13. All strainers, filters, and liquid-level glasses should be cleaned and checked before reassembly.

14. With the cleaning operation completed, reassemble the various components and make all joints and connections leakproof.

15. Before replacing clean-out ports, coat the machine surfaces with litharge-glycerol or ethylene glycol. Examine the gaskets for breaks or blistering and replace them if necessary.

16. When the unit has been reassembled, fill the various compartments to their normal operating levels with clean solvent and start the unit.

17. All controls and regulators should be checked and reset if necessary. The degreaser is then ready for operation.

11.4.6 Contamination measurement

One of the most important factors in the reliability of electronic circuits is their electrical stability with time. The presence of ionic contamination on insulating surfaces can seriously affect the electrical parameters, particularly in the presence of moisture. In addition, ions can cause corrosion of metallic conductors and a significant drop in insulation resistance between conductors over a short or extended period of time. When present, nonpolar or organic contamination is not as serious a problem unless it acts as a insulative barrier leading to false electrical testing results or masks over ionic contamination. However, measurement of nonpolar contaminants should still be a concern.

In manufacturing components, PCBs, and SMAs, corrosive ionic materials are generally used at one or more steps in the process. For instance, PCBs are plated, etched, handled by operators in assembly, coated with corrosive fluxes, and finally soldered. Of course, there is cleaning at various steps along the way, but, nevertheless, each step represents a potential source of ionic contamination which may be carried over onto the surfaces of the finished assembly. It is therefore most important to achieve a thorough decontamination of the surfaces after soldering.

There is a need to evaluate and measure the postsoldering cleaning process to determine the effectiveness of ionic residue removal. Measuring the resistivity of solvent extract is a common method of calculating residue ionic contamination on assemblies that have been cleaned. This method has been incorporated in several military specifications including MIL-P-28809, DOD-STD-2000-1, MIL-P-46843, MIL-P-55110, and WS-6536. Figure 11.30 compares the requirements of these specifications except MIL-P-55110, which is used for PCB fabrication.

Each specification uses the same basic test method initially developed by the Naval Air Development Center for testing assembly contamination prior to conformal coating. The test was primarily designed to confirm that all exposed areas (this does not include underneath SMDs) were sufficiently clean to promote effective conformal coat adhesion. Over time, its use has been carried over to testing assemblies.

This simple test method is well defined in MIL-P-28809. The test involves hand spraying a test solution of 25 percent isopropanol and 25 percent deionized water (by volume) from a wash bottle onto the assembly. The test solution is then collected and measured using a resistivity-conductivity bridge.

Of course this manual procedure has its drawbacks because of its operator dependency and its inability to remove only what can be considered a small portion of the residue's ionic contamination. It is apparent that contamination removal underneath SMDs would be next to nothing. Due to this limitation, instruments have been developed that can substantially reduce operator variables and can quantitatively measure the ionic contamination levels on assemblies.

Figure 11.30 Comparisons of Requirements Specified in Various Military Specifications for Assembly Resistivity of Solvent Extract Measurement

Specification	Test solution resistivity	Area calculation	Wash time	Lot formation	Sample size	Manufacturing point	Acceptable criteria
MIL-P-28809	6×10^6 cm min. acceptable; 25×10 cm expected through new deionizer	Surface area, both sizes, and estimate of components	1 min	Production of single shift	5 samples/shift	Before applying coating	$\geq 2 \times 10^6$ cm or equiv., μg NaCl/in^2
DOD-STD-2000-1	$> 6 \times 10^6$ cm, use deionizer, Hobson method; $>20 \times 10$ cm, instrumental method	Surface area both sides	No requirement	No requirement	MIL-P-105; AQL 1.0, S-2	Postsolder cleaning	$\geq 2 \times 10^6$ cm or equiv., μg NaCl/in^2
MIL-P-46843	6×10^6 cm min.; no deionizer requirement	Surface area both sides	No requirement	Identify and trace through soldering and cleaning operations	MIL-P-105; AQL 2.5, S-4	Postsolder cleaning	$\geq 2 \times 10^6$ cm
WS-6536	Per MIL-P-28809	Surface area both sides, and area of components	5 min	Restriction on variety in lot	MIL-P-105; AQL 1.0, S-2 min.	Postsolder cleaning and prior to coating	$\geq 2 \times 10^6$ cm or equiv., μg NaCl/in^2

11.4.6.1 Instruments for ionic contamination measurement. There are several instruments that are commercially available to measure the quantitative levels of ionic contamination on assemblies. Figure 11.31 lists these instruments, their manufacturers, and their MIL-P-28809 acceptance limit.

As explained by MIL-P-28809, the method of testing is founded on the principle that an isopropanol test solution of isopropanol and deionized water has a very low electrical conductivity. The assembly to be tested is immersed in the test solution contained in one or more tanks. The mixture dissolves the surface ionic contamination, which leads to an increase in solution conductivity. The change in conductivity recorded by the instrument is a measure of the amount of ionic residues dissolved in the solution. The conductivity of the solution is a linear function of the concentration of dissolved ionics and is, therefore, much easier to interpret than resistivity. Each instrument is capable of automatically calculating the level of contamination and provides each reading in micrograms of sodium chloride equivalents per square inch (μm NaCl/in^2). Although each type of instrument has different levels of sensitivity and accuracy, an instrument typically consists of a metering pump, a mixed-bed deionizer column, a test chamber(s) or tank(s), an in-line conductivity cell, a meter, an integrator, a magnetic stirrer, a plotter, and a control console or computer. The schematic illustrated in Fig. 11.32 shows a typical configuration of a conductivity solvent extract instrument. With this type of instrument design, the solution in all three tanks is circulated sequentially until a preselected conductivity baseline is obtained. The test tanks are isolated from the loop. A test assembly is placed in one of the tanks and allowed to extract for a set period of time. The test assembly is removed from the tank and then the test solution is pumped through the measuring section until the preselected conductivity baseline is reached. The integrator then displays the sodium chloride equivalent of the ionic contaminants.

Figure 11.31   The Various Types of Instruments Commercially Available to Measure Ionic Contamination on SMAs

Instrument	Manufacturer	MIL-P-28809 acceptance limit, μg NaCl/in^2
Contaminometer	Protonique	—
Ion Chaser	Dupont	32.0
Ionograph	Alpha Metals	20.0
Omega Meter	Kinco Industries	14.0

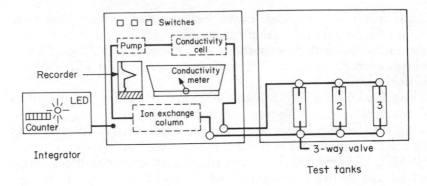

Figure 11.32   Schematic of a conductivity solvent extract instrument.

11.4.6.2 Drawbacks with ionic contamination measurement instruments.
The ionic measuring instruments discussed serve a useful purpose in a
more comparative analysis of cleaning process control. These instruments can also be used to compare the effects of different types of solvents, fluxes, and solder pastes, which the process engineer may want
to incorporate into manufacturing but does not yet understand the
cleaning effects the material change will have. They have also been
used to test components and PCBs.

With the many useful purposes these ionic contamination measuring instruments can have, there is at least one major problem currently plaguing the industry. With their current designs, the commercially available instruments do not provide adequate removal of flux
residues (especially rosin flux residues) underneath SMDs. The main
reason for this is that they do not promote sufficient amounts of mechanical action or exchange cycles of the test solution underneath the
SMDs. Furthermore, the solubility of rosin flux residues underneath
SMDs has been reported to be exceptionally poor with the current test
solution of 75 percent isopropanol and 25 percent deionized water as
specified by MIL-P-28809.[9] This, of course, indicates that checking the
cleanliness of SMAs with MIL-P-28809–qualified instrumentation
gives no indication of what ionic contamination lies underneath the
components.

What is required for these instruments to work more efficiently is
simply mechanical action such as spray, ultrasonics (if made acceptable), or a combination of both, and definitely a better performing test
solution. The companies that manufacture these instruments understand these drawbacks and are currently investigating their resolution.

11.4.6.3 Methods for measuring nonionic contamination. Nonionic contaminants or organic contaminants are not as serious a concern as

ionic contaminants; however, in some instances, they can cause some type of mechanical or electrical failure. This can consist of interfering with good conformal coating and solder mask adhesion for mechanical concerns and acting as an electrical insulator between the test land and test probe.

Currently there are no automated instruments commercially available that can measure nonionic contaminants. However, there are some manual test methods developed by IPC and Dow Chemical.

IPC has two methods as specified and described in IPC-TM-650, test methods 2.3.38 and 2.3.39. Method 2.3.38 is considered an "in-house method" and will neither identify the contaminants present nor separate contaminant mixtures into individual constituents. Method 2.3.39 uses infrared spectrophotometry for its determination of nonionic contaminants. The test is specifically designed to confirm the presence of and to identify nonionic contaminants of the arylakyl polyether family, but is not limited to contaminants containing an ether linkage. These test methods can be obtained from IPC at a minimal cost.

W. F. Richey and other colleagues of Dow Chemical have recently developed an ultraviolet absorption spectrophotometric analysis test method designed primarily to determine quantitatively the residue of rosin remaining on the assembly. Since there are currently no specifications available describing this test method, it is provided as follows:

Spectrophotometric determination of residual rosin on defluxed circuit assemblies (patent applied for)

Extraction step

Step 1: Place an accurately measured volume (using pipet, repipet, or syringe) of high-performance-liquid chromatographic-grade or spectral-grade isopropanol into each of two antistatic plastic bags. Pure isopropanol or aqueous isopropanol with up to 25 volume percent deionized water may be used. Choose an initial volume in cubic centimeters equal to three times the surface area in inches of the circuit assembly to be tested.

Step 2: Place the defluxed circuit assembly into one of the two bags containing isopropanol.

Step 3: Seal the zipper closures on the bags and fold the bags as needed to maximize contact between solvent and circuit board surfaces.

Step 4: Shake both bags by hand, or by mechanical shaker if the bags offer EMI/RFI protection, for 10 min.

Step 5: Remove circuit assembly from its bag.

Step 6: Pour the contents of each bag into labeled glass sample bottles, with tightly closed screw caps, for storage until ready to make the UV absorbance measurements.

Absorbance measurement (242 nm)

Step 1: Follow the spectrophotometer manufacturer's instructions for operating the instrument. Calibrate it with the quartz sample cell in place to allow the instrument to compensate for absorbance by the cell.

Step 2: Set the wavelength at 242 nm.

Step 3: Set the absorbance range at the maximum for the instrument.

Step 4: While holding the quartz sample cell by the frosted surfaces, use a disposable glass pipet to carefully rinse the cell with the solution to be measured as follows: Rinse the inside walls of the cell with solution stream from the pipet, filling cell to about one-third level, then remove all of the rinse solution from the cell with the pipet and discard the solution. Repeat rinse procedure three times, being careful to remove *all* of the rinse solution each time.

Step 5: After the sample cell has been thoroughly rinsed as described in step 4 above, fill the cell with the solution to be measured. Place filled cell in sample compartment of the instrument, close sample door, and read absorbance.

Step 6: The solution from the bag in which the circuit assembly was shaken is the *sample;* the other solution is the *blank.*

Step 7: If the absorbance for the sample is the same as that of the blank, repeat the entire procedure, but use another circuit assembly and less volume of isopropanol. Absorbance is linear with concentration of absorbing species, so one may estimate the volume required to obtain an absorbance reading in the range 0.4 to 0.6, where the instrument is the most responsive.

Step 8: If the absorbance for the sample is greater than 1.4, repeat the procedure, but use another circuit assembly and a greater volume of isopropanol (see step 7 above).

Calculation of residual rosin ($\mu g/in^2$)

Step 1: Subtract the absorbance reading obtained from the blank from that of the sample.

Step 2: Using the calibration curve, find the concentration of residual rosin (ppm rosin) correlating with the absorbance difference obtained in step 1.

Step 3: Substitute the appropriate numerical values into the following equation:

μg rosin/in^2

$$= \frac{\text{ppm rosin} \times \text{solvent density, g/cm}^3 \times \text{solvent volume, cm}^3}{\text{board surface area, in}^2}$$

$$\text{Density of 100\% isopropanol} = 0.785 \text{ g/cm}^3$$

$$\text{Density of 75\% isopropanol/water} = 0.85 \text{ g/cm}^3$$

Caution: Use proper handling and safety precautions while conducting this analysis.

Calibration procedure for spectrophotometric analysis
Stock solution preparation

Step 1: Weigh exactly 0.0785 g gum rosin or flux solids: (1) directly into a 1-L volumetric flask, or (2) into a smaller container, such as a beaker, then quantitatively transfer the rosin using high-performance liquid chromatographic (HPLC) or spectral-grade isopropanol into the 1-L volumetric flask. Rinse the beaker with the isopropanol at least 10 times, pouring all the rinse solutions into the volumetric flask.

Step 2: Fill the flask to within 1 in. of the calibration line on its neck.

Step 3: Shake the stoppered flask vigorously, turning it upside down several times to dissolve the rosin and to thoroughly mix the solution.

Step 4: Let the flask sit for about 30 min to allow liquid to drain from the neck. Bring the volume to the 1-L mark by adding isopropanol dropwise. This is the stock 100 ppm rosin solution from which dilutions are prepared for constructing the calibration curve. Keep solutions tightly stoppered to prevent evaporation.

Dilutions

Step 1: Using an appropriately sized volumetric pipet, transfer exactly the volume of the stock solution indicated in the table below into a 100-mL flask for each dilute solution desired:

ppm rosin in final (dilute) solution	1	2	5	10	20	50
mL stock solution	1	2	5	10	20	50

Step 2: Add HPLC isopropanol to each flask to a level within 0.5 in. of the calibration mark. Shake thoroughly before filling to the mark following the same technique outlined above in steps 3 and 4.

Calibration curve construction

Step 1: Follow the procedure for absorbance measurements (242 nm). The blank is HPLC isopropanol from the reagent bottle.

Step 2: Plot the absorbance difference (sample-blank) versus ppm rosin. The points should fall on a straight line which passes through zero.

Step 3: Points for concentrations greater than 50 ppm rosin may lie slightly below the calibration line, but these concentrations are unlikely to be encountered.

References

1. Carmen Capillo, "Surface Mount Assemblies Create New Cleaning Challenges," *Electronic Packaging and Production,* August 1984.
2. Rudolph F. Klima and Hillel Magid, "Anaylsis of Fluxes and Extracts of Printed Wiring by High Performance Liquid Chromatography," *Nepcon West Proceeding*s, February 24, 1987.
3. David G. Lovering, "Rosin Acids React to Form Tan Residues," *Electronic Packaging and Production,* February 1985.
4. Tim D. Cabelka, Wesley L. Archer, and Joseph A. Trombka, "Cleaning SMCs with Chlorinated Solvents," *IEPS SMT,* vol. 3, Part 2, 1986.
5. R. P. Musselman and T. W. Yarbrough, "The Fluid Dynamics of Cleaning under Surface Mounted PWAs and Hybrids," *IEPS SMT,* vol. 3, part 2, 1986.
6. C. K. Ellenberger, "TI Finds Best Cleaner, Parts I and II," *Circuits Manufacturing,* May 1986.
7. D. R. Gerard, "Eight Different Cycles for Cleaning Surface Mount Assemblies," *Circuits Manufacturing,* March 1987.
8. R. E. Martz, et al., "Materials Research Report 3-78," Naval Avionics Center, Indianapolis, Indiana, August 29, 1978.
9. J. E. Hale and W. R. Steinacker, "Complete Cleaning of Surface Mount Assemblies," *Nepcon West Proceedings,* February 26, 1985.

Bibliography

Adamson, A. W., *The Physical Chemistry of Surfaces,* John Wiley and Sons, New York, 1967.
Barton, A. F. M., *Handbook of Painted Technology,* Charles Griffin and Company, Ltd., London, England, 1969.
Bonner, J. K., " A New Process for Cleaning Surface Mount Assemblies," *Nepcon West Proceedings,* Cahners Exposition Group, February 24, 1987.
Breunsbach, R., "Batch Aqueous Cleaning: An Overview," *Nepcon West Proceedings,* Cahners Exposition Group, February 24, 1987.
Brous, J., "Evaluation of Post Soldering Flux Removal," *Welding Journal Research Supplement,* December 1975.
Dunn, D. B. and Bergendahl, C. G., "An Evaluation of the Solvent Extraction Method

for the Detection of Ionic Contamination on Substrates Supporting Large Surface Mounted Devices," *Brazing and Soldering,* no.13, Autumn 1987.

Ellis, B.N., *Cleaning and Contamination of Electronics, Components, and Assemblies,* Electrochemical Publications Limited, 1986.

Hayes, Michael E., "Cleaning SMT Assemblies without Halogenated Solvents," *Surface Mount Technology,* December 1988.

Kenyon, W. G., "How to Use the Solvent Extract Method to Measure Ionic Contamination of Printed Wiring Assemblies," *Insulation Circuits,* March 1981.

Manko, H., *Soldering Handbook for Printed Circuits and Surface Mounting,* Van Nostrand Reinhold Company, New York, 1986.

Morgans, W. H., *Outlines of Paint Technology,* Charles Griffin and Company, Ltd., London, 1969.

Tautscher, C. J., *The Contamination of Printed Wiring Boards and Assemblies,* Omega Scientific Services, 1967.

Willis, R., "SMD Cleaning: A Practical Assessment," *Printed Circuit Assembly,* October 1987.

Testing SMAs

Introduction

SMT provides the design engineer with many advantages such as the higher packaging density possible with substantially smaller SMDs and the ability to mount these components on both sides of the PCB. But to the test engineer, these so-called advantages can produce insurmountable hurdles in the ability to adequately test an assembly containing SMDs. The more SMDs there are on an assembly, the greater the functional density and the more inaccessible the test lands. As a result, the shift from conventional assemblies to SMAs has become a major concern for test engineers and manufacturers of PCB testers. In their quest to overcome the hurdles imposed by SMAs, test engineers must adequately understand PCB design rules for testing and take more seriously the need to pretest SMDs and the bare PCB prior to any assembly.

Another approach to consider is not to test at all, with the exception of functional testing used as a go–no go test. This approach, which can be considered the ultimate goal for any product, can be started from the standpoint of testing certain complex SMDs, bare board testing, and in-circuit testing of the assemblies. In all levels of testing, the goal is to work with the supplier and assembler to continuously reduce the defects found at testing to a point where testing is no longer economical. Unless test is required by the customer to prove product conformance, test only serves as a crutch in identifying problems created by the manufacturing process.

In-Circuit and Functional Testing

12.1 Assembly Faults

Assembly faults detectable at in-circuit testing can derive from multiple sources. The important factor regarding assembly faults is to recognize the location, the type, and the frequency of the faults detected. Keeping records and tracking these faults is paramount to reducing and finally eliminating these faults.

Assembly faults can be broken down into five different groups: PCB fabrication faults, soldering faults, assembly errors, defective components, and functional failures. Figure 12.1 illustrates the percentages of each of these assembly faults commonly found on tested assemblies. Typically, soldering problems account for half of the PCB faults, while assembly errors account for one-third. Faulty components and functional failures are the least frequent forms of defect found during testing.

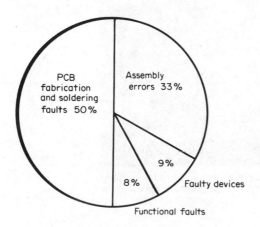

Figure 12.1 A typical mix of assembly faults and their percentages found at testing.

12.1.1   SMD faults

Whether conventional or surface mount, ICs are usually electronically tested at the die level and packaging level stages. This accounts for the normally low defect levels found with surface mount IC packages at the assembly level. Inoperative or partially operative ICs are still possible and can appear faulty during the manufacturing stages. Since the soldering process can impose sufficient amounts of thermal stress on the IC to cause problems, it is possible for pathways from the outside environment to develop between the lead frame surface and the component's packaging material. This in turn allows moisture and contaminants to make their way to the die, leading to component failure from contamination or corrosion. More common types of faults found with surface mount ICs are lead coplanarity package dimensions, damaged leads, and, worst of all, lead solderability. These component conditions leading to assembly faults can all be tested visually at incoming inspection, which is highly recommended for all types of SMDs.

Faults commonly found in passive SMDs such as chip resistors and capacitors include solderability of the end caps and internal construction faults (especially in capacitors) such as cracking and excessively porous ceramic. Figure 12.2 illustrates these faults, which are more common than expected, in a monolithiic chip capacitor. Again, per-

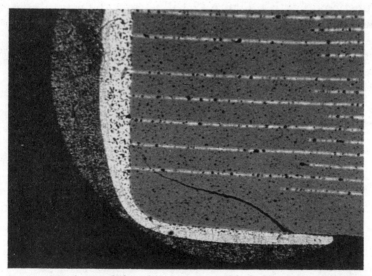

Figure 12.2   A monolithic chip capacitor with faulty ceramic material causing component cracking during thermal shock testing. Excessive porosity in the electrodes is also present.

forming incoming inspection tests of such factors as thermal shock, solderability, and internal destruction analysis on sample lot bases for chip capacitors is highly recommended. Chip resistors usually require only solderability testings.

12.1.2 PCB faults

PCB faults formed during the fabrication process can be quite numerous. Some of the more common faults consist of shorts between adjacent conductors; broken or open conductors, solder mask on lands, and plated through-hole faults such as layer-to-layer conductor misregistration, insufficient resin smear removal, and barrel cracking. When faulty PCBs find their way to the assembly level, repair of these faults is not usually possible, which can cause a substantial financial loss once the components are mounted and soldered onto the PCB. Since SMT is forcing designers to increase functional densities and therefore PCB fabrication densities, bare board testing will become more economically justifiable. Electrical testing of bare PCBs normally consists of opens and shorts. When multilayer boards are used, additional testing should be performed to determine the quality of the PTH construction. This should include solder float testing of the coupons and their microsectioning analysis to determine conformance to the specified requirements.

12.1.3 Assembly faults

Assembly and soldering faults contribute to the majority of all faults found at testing. The most common types of faults are solder opens and shorts. These faults are attributed to insufficient manufacturing processes, but sometimes they occur due to the faults passed on from PCB design, SMDs, and PCB fabrication. Solder opens can occur due to component misalignment, insufficient printing of solder paste, and a variety of process insufficiencies when wave soldering is used. In most cases, opens are caused by insufficient process controls and component solderability. Solder shorts are nearly always caused by a lack of process controls.

Other faults such as wrong component locations and missing components can also occur during component placement and sometimes during the wave soldering process, respectively. These faults occur simply because of the lack of proper process controls and of proper quality feedback while assemblies are processed. These kind of faults are the simplest to correct and should only be a temporary condition if they do occur.

12.2 Types of Testing[1]

12.2.1 In-circuit testing—analog

Analog in-circuit testing uses measuring operational amplifiers (MOA) to isolate and measure components. Figure 12.3 illustrates the basic circuit for a MOA. The unknown impedance Z is stimulated by a known voltage (originating in the board testing system) at node S. Node I is connected to the inverting input of the MOA and, through operational amplifier action, becomes a virtual ground. The same current i that is forced through Z also flows through the MOA's feedback resistor R_{ref}. R_{ref} can be computed from the voltage at the output of the MOA, shown as V_O. The analog-to-digital (A-to-D) converter allows the test system's digital circuitry to measure analog voltages. Even though the MOA power supply voltage V_S may limit the range of V_O, R_{ref} can be changed in order to measure a wide range of components.

This basic technique can be applied to all passive components. For reactive components, V_S can be changed from a dc source to an ac source and detection made by an ac detector or phase synchronous detector instead of a dc detector.

In many cases, the simple MOA approach will not work because of parallel paths around the unknown component formed by other PCB components. In Fig. 12.4, this problem is solved by using guarding. By connecting node G to a guard, node I is at the same potential (virtual ground), so no error current will flow through Z_{P2}. The source voltage V_S does supply current through Z_P, but a voltage drop across Z_P does not affect the measurement (provided the source impedance of V_S is low).

This simple form of guarding has some imperfections in an actual board testing system. Since various nodes are multiplexed in the test-

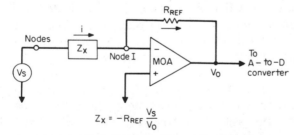

Figure 12.3 The basis of analog in-circuit testing is a measuring operational amplifier (MOA). A known voltage V_S is applied to the component being tested (Z_X). By operational amplifier action, the ouput voltage is proportional to i and R_{ref}. Using the known values of V_S and R_{ref}, the value of Z_X can be determined.

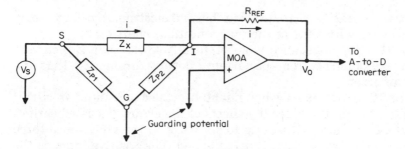

Figure 12.4 Parallel paths around the component to be measured introduce significant errors. By using a guarding technique, these errors can be reduced. The general idea is to shunt current through Z_{P1} to guard G and to prevent current from flowing in Z_{P2}.

ing system through relays and cable runs, system resistances and thermal offsets introduce measurement errors as shown in Fig. 12.5. Manufacturers of board testing systems attempt to reduce the effects of testing system–induced errors; however, their effects can still influence measurement accuracy.

One type of error encountered is the source voltage V_S caused by source resistance R_S between the source and unknown. V_S is used to calculate Z_X, but the unwanted voltage drop across resistance R_S and lead resistance R_l will produce an error, causing Z_X to appear to be larger than it actually is. If Z_{P1} happens to be small, then an error current i_1 will be dropped across R_S.

Another potential source of error is in the guard circuit itself. For guarding to be effective, positive MOA input must be held at the same potential as node G. Any offset voltage E_{OS} or current flowing through lead wire resistance R_O will cause node G to float at a potential not equal to the positive input of the MOA and thus appears as an error in

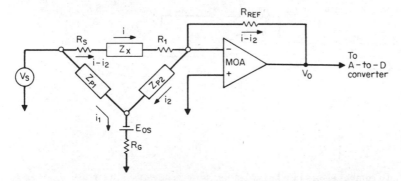

Figure 12.5 Simple guarding has its imperfections when applied to an actual board testing system. Resistance through cable runs and terminal offsets introduces measurement errors.

V_O. Worse yet, if E_{OS} is created by a thermal electromotive force, readings will vary with time as thermal conditions change.

When the Z_{P2} impedance is substantially lower than Z_X, current splitting will subtract from the current i flowing through Z_X, thus producing an error.

Figure 12.6 illustrates what might be called advanced in-circuit testing to attempt to reduce the effects of the errors discussed earlier. This extended guarding uses up to six wires to remotely sense three additional nodes of the circuit being tested. The positive input to the MOA can be connected directly to node G, which reduces the unwanted effects of R_G and E_{OS}. By bringing the feedback path through R_{ref} out of the board testing system to node I, the effect of R_I can be virtually eliminated (there will still be some lead resistance, R_B). This approach will reduce errors produced by source resistance R_S and current i_2 flowing through Z_{P2}. By using the A-to-D converter directly across R_{ref}, the effect of lead resistance R_B is reduced (R_B is usually much smaller than R_{ref} thus contributing little to an error in voltage drop across R_{ref}).

A second element in extended guarding involves measuring the voltage directly across the unknown Z_X rather than measuring V_O. To do this, the voltage at node S is sensed on one side of the A-to-D converter, and the negative input to the MOA is sensed on the other side. The effect of lead resistance R_I is eliminated due to the high input impedance to the MOA.

Some board testing systems use accuracy enhancement to negate other nonideal system characteristics by making a series of internal measurements and calibrations within the system. This improves re-

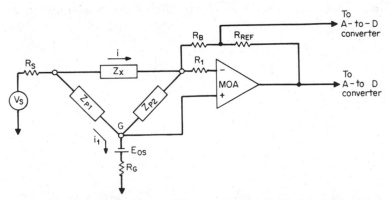

Figure 12.6 To combat errors in a board testing system, advanced in-circuit testing is used to extend guarding by using six wires to remotely sense additional nodes on the circuit being tested.

peatability by canceling the effects of all system offsets as well as time- and temperature drift–induced variations.

Nothing is free; i.e., advanced in-circuit testing requires up to three additional wires per tested component for remote sensing. This translates into more scanner relays (used to sequentially move board testing connections from component to component) and higher fixture costs. Improvements in accuracy using these approaches can sometimes justify the additional expense, but longer measurement items are required.

The approach used to measure the value of a capacitor is similar to that used for a resistor except that an ac source is used for V_S. The ac source should have selectable frequencies (typically 100 Hz, 1 kHz, and 10 kHz) to cover different capacitance values. The simple, unguarded approach shown in Fig. 12.7 may be all that is necessary for capacitors without parallel circuit paths; an ac detector measures the value of V_O. Care must be taken to keep V_S as low as practical to prevent turning on any semiconductors in the circuit being tested.

As is true with resistance measurements, parallel paths around capacitors cause errors that can be reduced by using guarding techniques. A challenging in-circuit test problem is shown in Fig. 12.8 where the capacitor has a small parallel resistance around it. A phase-synchronous detector is used to resolve real and imaginary components of the impedance.

Accurate in-circuit measurements of inductors are complicated by their inherent series resistance, as illustrated in Fig. 12.9. At test frequencies below the megahertz range, resistive impedance R_L can be quite large relative to inductive reactance. Again, a phase-synchronous detector makes it possible for the board testing system to separate the effect of series resistance from inductance.

Various types of semiconductors can also be measured in circuit using the MOA. Take, for example, the simple diode shown in Fig. 12.10. It is placed in the MOA feedback loop in such a way that it is forward-

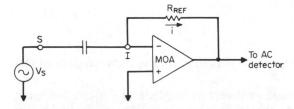

Figure 12.7 The approach used to measure the value of a capacitor in-circuit is similar to that used for a resistor except that an ac source replaces the dc source.

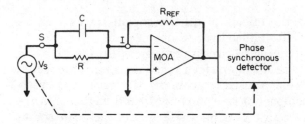

Figure 12.8 A measurement challenge to in-circuit testing occurs when the capacitor has a small parallel resistance. A phase-synchronous detector is used to separate the real and imaginary components of the impedance.

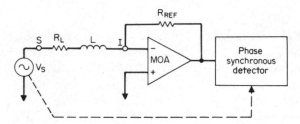

Figure 12.9 Inductors have an inherent series resistance, and a phase-synchronous detector is employed to separate the effect of the series resistance from the inductance.

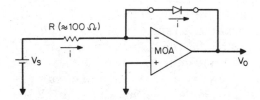

Figure 12.10. The measurement of a simple diode involves placing it in the feedback loop of the MOA so that it is forward biased. The output voltage, V_O, equals the forward drop across the diode.

biased. The output voltage V_O is a reflection of the forward drop of the diode.

As a side note, a misloaded or defective diode in the MOA feedback loop can cause V_O to rise to the MOA's supply voltage and potentially damage the circuit being tested. Many in-circuit testing systems therefore make provisions to clamp the maximum possible MOA out-

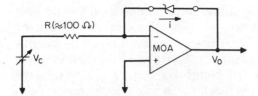

Figure 12.11 Zener diodes are tested in-circuit with V_O equal to their breakdown voltage. These tests are limited, however, to diodes with breakdown voltages equal to or less than the MOA's supply voltage.

put voltage just slightly above the expected forward voltage drop of the diode being tested.

Zener diodes are tested in much the same way as other diodes except that the voltage polarity is reversed in such a way that the MOA's output equals the zener breakdown voltage. The test is, however, limited by the power supply voltage to the MOA (often less than 15 V). Testing higher voltage zeners may have to be done at incoming inspection or deferred to final testing and assembly. The MOA configuration for testing a zener diode is shown in Fig. 12.11.

Bipolar transistors can be tested using the configuration shown in Fig. 12.12. The transistor is placed in the MOA's feedback loop and a dc bias current and an ac signal are injected into the base. The dc bias must be sufficient to turn on the base-emitter junction, and resistors R_1 and R_2 must be of the proper value to maintain a forward base-

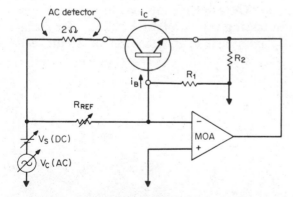

Figure 12.12 Bipolar transistors are tested under biased conditions using a dc bias voltage superimposed on an ac signal. An ac detector is used to sense the voltage across the 2-Ω resistor to compute the transistor's gain.

emitter junction voltage. The ac signal superimposed on the dc bias will be amplified by the transition gain. The ac signal can be sensed across the 2-Ω collector resistor and the gain measured. The magnitude of the dc bias, ac signal, and bias resistors can be varied to allow testing of different components. Although far from a comprehensive parametric test, this in-circuit approach to testing transistors will detect most faults by determining if the component is "alive" and can amplify small signals.

Field effect transistors (FETs) can be tested by placing the component in series with the MOA negative input. As shown in Fig. 12.13, this configuration can verify $R_{D(ON)}$ and I_{DSS}. By rearranging this circuit and allowing the MOA output to drive the gate, FET pinchoff voltage is tested, as shown in Fig. 12.14. The drain-to-source voltage is established by V_S causing some drain current to flow into the MOA's summing node I (a virtual ground). This circuit will stabilize itself at a drain current close to I_{DSS}, and V_O will closely approximate the FET's pinchoff voltage.

In summary, analog in-circuit testing makes extensive use of a MOA to test a variety of both passive and active components. Errors caused by parallel impedance paths around components being tested can be reduced by using guarding. In-circuit testing typically uses a bed-of-nails test fixture for access to any given node on the board under test. Because each component is tested against at least a partial set of its parameters, fault visibility is high. Software investment for in-circuit testing is typically lower than for functional testing, and in-circuit testing is noted for finding the largest number of faults for the least investment. The PCB operation is not, however, tested for its ability to perform functionally. Testing time on good boards also may not be appreciably less than testing time on faulty boards. For situations where there are high percentages of good boards, in-circuit test-

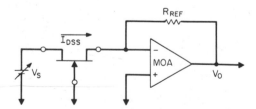

Figure 12.13   FETs can be tested for $R_{D(ON)}$ and I_{DSS} by placing the component in series with the MOA's negative input. V_S is programmed to force the FET to operate in its linear region allowing $R_{D(ON)}$ to be measured.

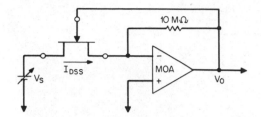

Figure 12.14 By allowing the output of the MOA to drive the gate of a FET, its pinchoff voltage is reflected in the magnitude of V_O. This comes about as a result of the negative voltage V_O fed back to the FET's gate, which causes the circuit to stabilize at the pinchoff voltage.

ing usually ends up being slower than functional testing. This is especially true as PCB complexity increases.

12.2.2 In-circuit testing—digital

In general, digital in-circuit testing also attempts to electrically isolate and test each component on a PCB, thus providing high fault visibility. Digital components are tested according to the specifications embodied in their truth tables. A digital in-circuit testing system typically includes a library of truth tables or other information needed to make up the stimulus for a given component.

All digital components may be placed in two general categories: (1) combinational components that have no memory of previous events, and (2) sequential components that have memory and whose next state depends on previous states. A NAND gate, for example, is combinational. Its output is determined solely by its present inputs. In contrast, the outputs of a sequential component (such as a microprocessor) depend not only on its present inputs but also on previous inputs. This distinction is of great significance in testing digital components.

The digital logic levels present at the inputs and outputs of a component are referred to as states. Some possible states represent proper behavior while others represent defective behavior. For combinational components, there is no ambiguity as to which states are correct and which are not. Present input always determines present output. In contrast, knowledge of the present input to a sequential component does not allow its output to be predicted because the sequence leading up to the present state must also be known. Figure 12.15 shows examples of both combinational and sequential components and their associated truth tables. Note that although the input states are the same

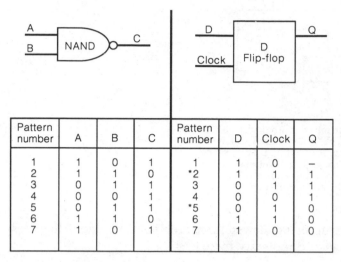

Pattern number	A	B	C	Pattern number	D	Clock	Q
1	1	0	1	1	1	0	–
2	1	1	0	*2	1	1	1
3	0	1	1	3	0	1	1
4	0	0	1	4	0	0	1
5	0	1	1	*5	0	1	0
6	1	1	0	6	1	1	0
7	1	0	1	7	1	0	0

* Positive clock transitions

Figure 12.15 A NAND gate and D flip-flop offer a contrast between combinational and sequential components. Note that for the same sequence of inputs, the output of the NAND gate can always be predicted. This is not so for the flip-flop where its output state depends on previous inputs.

for both components, the output of the NAND gate can always be predicted while the output of the flip-flop cannot.

The solution to testing sequential components is first to place the component in a known state to allow the output to be predicted or at least repeatable. Many components have a reset input which can be used to accomplish this. In some cases, it is necessary to use a specific series of input states (homing sequence) to return the component to a predictable output state.

At this point, it might be well to review some common failure mechanisms in digital components. A component may have been loaded in the wrong location or installed backward or some pins may have bent under the IC package thus creating an open circuit. The part could have been initially defective or perhaps damaged in handling by static electricity. Internal wire bonds that connect silicon chips to lead frames may be missing or broken. Output transistors within the IC could be shorted to the power supply or ground. There are also more exotic types of failures, but the above occur most frequently. The most common fault is the "stuck-at" model where the logic state of a node cannot be made to change state.

Nodes on the components of a PCB being tested are usually accessed through a bed-of-nails test fixture. The board being tested is pulled down onto the nails by vacuum. The nails in the custom test fixture provide intimate contact with both input and output pins of the components to be tested.

The basic digital in-circuit testing approach forces the input nodes of the component being tested to specific states. If enough current is injected into or withdrawn from a node, the natural state of that node may be momentarily forced high or low. A test sequence can then be literally forced into the component, thus allowing verification of its operation. Drivers are used for stimulation, and receivers are used to pick up the output of the component. In many cases, components can be tested in a pure "free space" sense as though independent of external circuitry. However, if a node is tied directly to a supply voltage or to ground, then node forcing will not work and such a topological restriction will prevent execution of the component's full truth table.

There are other topological restrictions that must be taken into account such as a feedback loop around the component to be tested. When the test system drives a node high, for example, a feedback loop may momentarily toggle the component's output node in a way that would indicate a failure. Another example occurs when testing a tristate driver with its output connected to a bidirectional bus. When looking at the output of the driver, traffic on the bus from other drives may produce false responses. As a general rule for in-circuit testing, the surrounding circuitry must be analyzed to see if it affects either the input or output of the component to be tested. In some cases, a feedback loop may have to be broken. In the case of testing tristate drivers, it may be necessary to inhibit traffic on the bus during testing.

Another approach to digital in-circuit stimulation uses a series of divide-by-N counters. Outputs from the divide-by-N counters are square waves. They have a power-of-two relationship which interacts to produce all possible combinational states over the period of the lowest frequency selected. This approach to in-circuit testing is used with a library of components which lists recommended frequencies to be applied to fully test that particular component.

The response to the stimulus is not predicted based on truth table information but is determined empirically. Signatures at the output pins based on the performance of a known good board are stored. All future boards to be tested are evaluated against these signatures.

A good example of this approach is illustrated in Fig. 12.16. For this D flip-flop, the highest divide-by-N frequency is applied to the clock input, and lower frequencies are assigned to asynchronous inputs such as PRESET and CLEAR. An intermediate frequency is assigned to the D input. The objective is to select frequencies that produce all possible combinations of input states during the period of the lowest frequency

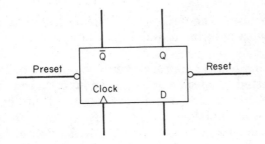

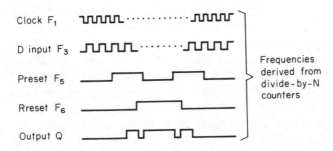

Clock F_1

D input F_3

Preset F_5

Rreset F_6

Output Q

Frequencies
derived from
divide-by-N
counters

Figure 12.16 This is an illustration of how the divide-by-N
approach to digital in-circuit testing can be applied to a D
flip-flop. Stimulation is accomplished by application of
square waves to the inputs of the component. The output is
recorded by the board testing system as a signature based on
the operation of a known good board. For the sake of completeness, however, the Q output is shown here for a D flip-
flop operating in free space.

selected. For this simple component, the goal could have been
achieved in various ways other than that illustrated in this example.

If the D flip-flop were isolated from surrounding circuitry, the Q
output would be shown. In actual practice, the output would not be
computed but rather stored as a signature allowing any topological restrictions to be ignored, simplifying programming.

The output data stream characterizes a complete truth table test for
this component. This output can be compressed into a unique hexadecimal signature dependent not only on the truth table response, but
also the time window over which the signature is recorded. Signature
analysis is used as a detection scheme to simplify storage of responses
without sacrificing the completeness of the test.

For sequential components, the divide-by-N approach has some limitations. Sequences composed of an instruction set become difficult to
set up. To some degree, this problem is reduced in severity by
software-intensive program generators which select appropriate frequencies to create subroutines. A library of the more popular sequen-

tial components is available with this type of testing system to supply what could be called test templates.

In contrast to the above method, another form of digital in-circuit testing involves taking control of the component being tested by using short-duration pulses instead of square waves. Verification is typically made by exercising inputs and checking outputs against the component's truth table. Verification can be limited to a partial check of the truth table covering only operations normal to the component's specific application or the full truth table for the component.

Testing is carried out dynamically by loading the test pattern into random access memory (RAM) and then applying the pattern at the speed at which the component normally operates. Responses are picked up on the output pins of the component by high-speed receivers and stored in RAM. This type of testing system is illustrated in Fig. 12.17. Tests for failures can be made on the fly as patterns are applied or after the testing is completed by analyzing the stored responses. Dynamic testing of this nature is essential on many modern digital components such as dynamic RAMs and microprocessors. In many cases, these components simply will not function at lower speeds.

The divide-by-N approach forces input nodes to change using square waves, which has a potentially damaging effect on the components due to excessive power dissipation within the component as well as other components connected to it. Potential damage is minimized by controlling pulse amplitude and width.

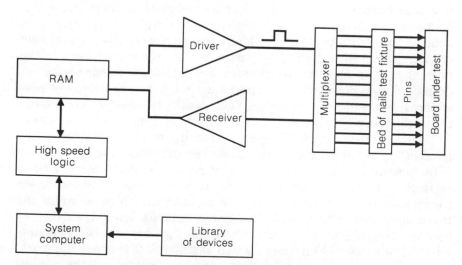

Figure 12.17 Short-duration pulsing is yet another approach to digital in-circuit testing. This approach recognizes the potentially damaging effect of node forcing on components; thus, pulse amplitude and duration are controlled.

There are some general rules to limit the amount of stress on the component being tested. If test patterns are of long duration, cooling cycles may have to be added to allow overdriven semiconductor junctions to return to ambient temperature. Also, as the number of nodes on a given component which must be overdriven increases, the maximum overdrive time must be proportionately lower. The voltage level at which overdrive takes place must also be taken into account. Its only necessary to overdrive a component just beyond its threshold level for detecting a high or low logic level on the output. One problem faced by manufacturers of automatic board testing systems relates to parasitic capacitance and inductance, primarily in the test fixture wiring, which can cause voltage overshoot or ringing to occur as pulses are applied. This obviously needs to be controlled.

Board testing systems invariably include some type of software package to aid in test development. When a test for a particular IC is called for, the test development program scans the component library to find the truth table or test sequence for that component. After determining which pins need to be driven and which need to be monitored, topological information then modifies the test. It should be obvious, for example, that a NAND gate with both inputs tied together (to form an inverter) cannot undergo testing of its full truth table.

Program generation becomes much more complex when applied to the testing of sequential components such as microprocessors. Using the component's RESET line or a homing sequence is often necessary to ensure that the testing begins in a known state. Attempting to test all possible instructions for a microprocessor will quickly drive test time up to unacceptable levels. Many test engineers use a limited set of instructions which might be part of a sequence encountered in the normal product operation. Some means of inhibiting traffic on the address and data buses which surround the microprocessor may be necessary to prevent false responses. It is usually necessary to break the topological loop between the microprocessor and its memory elements to isolate components in the loop. By looking at Fig. 12.18, the problems in testing sequential components become more obvious.

In summary, two approaches are currently available for in-circuit testing of digital components. One uses a series of divide-by-N counters to derive a series of square waves with which to stimulate the components to be tested. The output data stream is typically recorded as a signature characteristic of the component operating within the electrical environment on the board. The other approach uses short duration pulses designed to prevent damage to the components being tested. The stimulus is derived from the truth table for that specific component. The output is analyzed based on truth table information to determine if the component is good or bad.

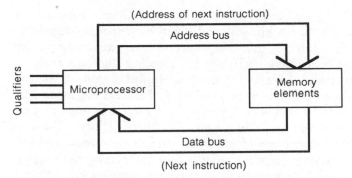

Figure 12.18   In microprocessor designs, a loop exists between the memory elements and the microprocessor itself. This inhibits fault isolation of the components within the loop and on the address and data buses. The loop usually needs to be broken (by removal of the microprocessor or by some built-in feature) before testing can take place.

The divide-by-N approach is limited in speed to the maximum frequency available from the counters. In contrast, an effort is made with pulsing to stimulate components at or close to normal operating speed. Keep in mind that certain sequential components cannot perform at clock rates appreciably below their normal clock rate. Both techniques offer the user high fault visibility leading to excellent diagnosis of potentially defective components.

12.2.3   Functional testing

Functional testing determines if an electronic product operates properly. Practically every product built today uses some form of functional testing, even if it is as simple as having a technician turn the product on and run it through a series of performance tests. Functional testing also can be applied to individual PC boards, and the simplest approach is the "hot mock-up" where the board tester is the product itself. Boards to be tested are first connected to the appropriate circuit within the product, where pressure is applied, and if the product works properly, then the PCB passes. This simple hot mock-up approach requires little investment but lacks the ability to automatically diagnose failures.

Functional testing is divided into static and dynamic testing. Static testing is generally lower in cost and is by far the most widely used test technique. The board being tested is observed in a fixed, nonchanging state. Dynamic testing stimulates the PCB, and activity is observed for proper operation at or near the clock rate under which the circuitry normally operates.

Either approach involves three basic elements: (1) the application of the stimulus, (2) the collection of responses, and (3) an evaluation of responses against either a known good board or responses computed for that board. If a fault is detected, most testers have a diagnostic routine or some means for identifying and locating the fault.

There are a number of ways to stimulate digital circuits. Test patterns, of course, can be generated manually. The test engineer analyzes the circuits to be tested in terms of likely failures and then designs a sequence of test patterns to discover these failures. Such an approach is labor intensive and requires meticulous record keeping. On large boards, manual testing quickly becomes too burdensome to manage. An automatic approach to stimulation involves the application of pseudo-random patterns or a counting sequence where one logic state is changed at a time. This approach attempts to cover all possible input combinations. Programming costs are relatively low, but an analysis is needed to determine if the input pattern provides adequate fault coverage. There are dangers in using this approach on microprocessors because applying invalid instructions can cause them to hang up. However, using a counting sequence to call up all possible addresses in a read-only memory (ROM) can be a very effective tool for fully exploring the contents of the component.

Using fault simulation is the most exacting approach for generating a stimulus. Based on the board's topology (entered by the test engineer) and a library of truth tables for the components used on the board, the simulator approach makes extensive use of a computer to create a model of the board being tested. The effectiveness of detecting specific kinds of faults can be tested for various stimuli. Since each node state on the board is computed, diagnostics can be developed to troubleshoot the PCB.

Many functional testers apply the stimulus and check the responses through the board's edge connector. Since the board is viewed from its edge connector, diagnosing failures on interior nodes is difficult. Many of these systems are equipped with a hand-held probe, thus allowing the user to access interior nodes. The simulator-equipped systems provide the user with directed probe placement instructions to back-trace to the failing node. Other functional board testing systems are equipped with a bed-of-nails fixture to automatically access interior nodes.

There are a variety of approaches to functional testing that differ widely in price and complexity. One approach, shown in Fig. 12.19, uses a dual fixture where a known good board is kept in one fixture and unknown boards are tested in the other fixture. The same stimulus (usually pseudo-random) is simultaneously applied to both boards, and responses are compared. This simple approach to functional testing depends on the board undergoing testing responding in precisely

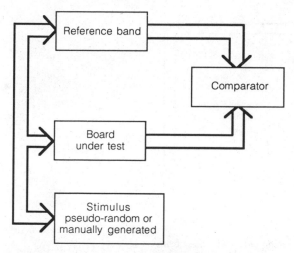

Figure 12.19   Reference board testing uses the response of a known good board as the basis for comparison with boards being tested. Software and hardware investment is modest for this approach.

the same manner as does the known good board. If a fault occurs, a hand-held probe is used to compare corresponding nodes and backtrace to the failing node. However, there is no guarantee that the test stimulus will uncover all possible faults by causing them to propagate to one of the output pins.

A similar approach with a single fixture first exercises a known good board and the stores the responses. From then on, all other boards are evaluated against the known board. This approach is illustrated in Fig. 12.20.

One problem with both of these techniques is that small timing differences between the known good board and successive boards tested can falsely indicate failures. Since the stimulus is typically pseudo-random, fault coverage is totally unknown. On the positive side, investment in these systems is low and very little software development is needed to make them operational. These systems usually use static testing where responses are observed in a fixed, nonchanging state.

Simulation is an exacting approach to functional board testing, but it is software-intensive and makes extensive use of the board testing system's computer. The responses of a physical circuit to sets of stimuli can be determined by using a software model rather than actually looking at the operation of the circuit. Inputs include the topology of the board to be tested and a library that characterizes the operation of the components used on the board. Correct responses to selected input patterns can be predicted. By calculating the logic state of each PCB

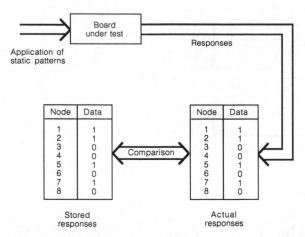

Figure 12.20 Stored-response testing derives its data from a known good board. Responses are stored in the system's computer, and all other boards are then compared with the known good board.

node, the effectiveness in finding specific kinds of faults can be evaluated. This allows the simulator to compute fault coverage for each proposed stimulus and thus aids in the development of the optimum stimulus. All information necessary to troubleshoot the board being tested is contained in the state file for each node. Many simulator-based functional testing systems include a hand-held probe, as shown in Fig. 12.21, thus allowing the operator to isolate a fault to the fail-

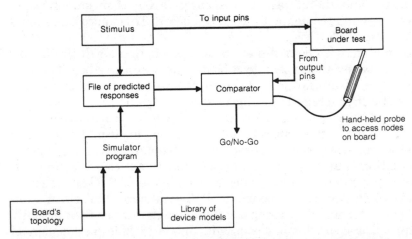

Figure 12.21 A simulator compares the response of a given stimulus to the predicted response based on a model of the circuit being tested. Although software-intensive, the simulator approach allows the user to compute fault coverage and to scientifically generate the stimulus.

ing node under the guidance of computer-generated instructions. Since the simulator models a given circuit, it could also double as a design tool (a point worth noting in financial justification proposals).

A simulator is only as good as the model, and it is impractical to build a model that duplicates a digital process in every detail. Models are, therefore, built to represent the most important characteristics of a circuit. The test engineer must know the model well enough to understand how it will respond to different stimuli. There are some significant limitations to modeling, i.e., for complex sequential components, such as microprocessors, a gate-level equivalent is next to impossible to originate. Manufacturers of such components are reluctant to provide the necessary information to develop a model and therefore may limit the type of circuitry for which simulator testing can be used.

The majority of simulator-based functional testers use static testing. There are some on the market that test boards dynamically; however, these systems are expensive since RAM-backed, high-speed drivers are necessary to apply the stimulus at high clock rates. In addition, RAM-backed receivers are necessary to pick up the responses. Faults of a dynamic nature, however, can be detected, and the rate at which testing can be performed is greatly improved.

12.2.4 Signature analysis testing

A less costly approach to dynamic digital testing uses a measurement technique called signature analysis (SA). A synchronous digital receiver is used to collect the data stream at the node being tested. SA compresses the data into a single, unique, hexadecimal signature. The data stream is observed during the preselected time window governed by start-stop signals selected by the test engineers. In this way, only data that are meaningful to a given test are observed. The board being tested is allowed to free-run or is stimulated using a repetitive pattern so that the data streams are repeated over and over. In this way, unstable signatures (created by parts that operate marginally) can be detected.

A bed-of-nails test fixture which gives access to any given node on the board being tested is necessary for SA testing. Proper operation of a given component can be viewed by examining the signatures on the inputs and then on the outputs. Fault isolation is conducted by using back-tracing methods in a manner similar to a technician using the oscilloscope to look at waveforms on an analog circuit.

Good signatures are derived from a known good board and are stored in the system's computer. As each node being tested is multiplexed by the board testing system, its signature is compared to that

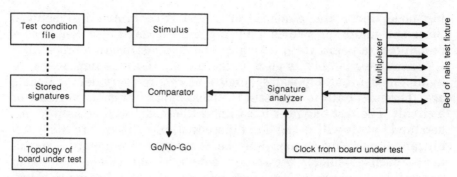

Figure 12.22 As shown in this block diagram of a board testing system using a signature analyzer, the signatures derived from a known board are compared with those of the board being tested. Testing is done dynamically with the signature analyzer synchronized to the board's clock.

derived from the known good board. The block diagram for a board testing system using SA is shown in Fig. 12.22.

From a hardware standpoint, a signature analyzer is a set of shift registers with carefully selected feedback taps as shown in Fig. 12.23. These taps minimize the probability of a bad node producing a good signature (on some units this probability is as low as 0.002 percent for the detection of multiple errors and 0 percent for single-bit errors.) The occurrence of selected levels at the START and STOP inputs initiates and terminates the measurement period or window. An example would be to begin a test sequence of a memory element (such as a ROM) at its starting address and then terminate the sequence at the ending address.

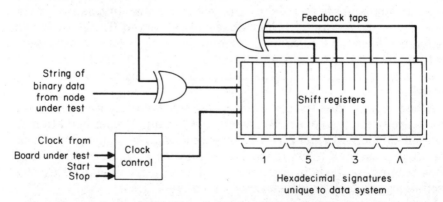

Figure 12.23 Signature analysis is a measurement technique used to compress long strings of binary data into a single unique signature during a preselected START-STOP window.

The signature itself is derived from the remainder left in the shift registers at the end of the start-stop window. The signature is then characteristic of the time-dependent logic activity for a particular node. Any change in mode behavior from the known good board to the one being tested will produce a different signature, thus indicating a possible timing difference, a skew in the clock, or a lost bit. Being stuck at high or low faults is easily detected by connecting the ana- lyzer to the supply voltage, recording the signature, and noting that a stuck low signature is composed of all zeros.

The test engineer must concentrate on generating a proper stimu- lus. It must propagate its way through the board being tested and re- veal failures as it goes. To accomplish this, it may be necessary to stimulate various areas of a board separately. In some cases, boards are designed to provide their own stimulus when actuated by putting the board into a free-run testing mode.

There are quite a few different approaches to functional testing, as discussed herein, and there are even more forms available on commer- cial testing systems. The situation should be viewed as a selection of tools rather than right ways or wrong ways. What works well in one case may not work at all in another case. Economics and throughput will quickly enter the picture as a potential user starts looking at the price tags. For example, fault coverage may not be as high using a tester which relies on stored responses, but an elaborate simulator- based tester may be too expensive for the yield involved. On the other hand, the probability of shipping a defective unit may dictate a more comprehensive approach to testing.

The reference board approach and stored response method of static pattern testing require only a modest investment in both hardware and software. Although simple to implement, they rely heavily on the known good board being representative of all future boards to be tested. These approaches also rely on the stimulus to reveal failures and offer little help in computing fault coverage. The pseudo-random stimulus so commonly used for these approaches will fall short of properly testing sequential components (such as microprocessors).

The simulation approach can be applied to either static or dynamic testing and precisely models the board being tested. In the process, the logic state of every node is predicted for a given stimulus. Fault cov- erage can be computed and faults can be located using a hand-held probe. Not all modern LSI components can be modeled, however. This approach is expensive and is both hardware- and software-intensive.

Signature analysis is a measurement technique for testing boards dynamically and does not require modeling. Software requirements are less than for simulation. The technique, however, relies on the known good board being representative of all future boards to be tested. The nature of the fault is not analyzed precisely (unless it is stuck at a high or low), and the only information the user gets is that the signature at a given node is bad.

12.3 Automatic Test Equipment

The term automatic test equipment (ATE) applies generally to all forms of automatic testing including those used in the manufacture of components, incoming inspection, PCB fabrication, and board assemblies. To begin with, manual testing is an alternative to automatic testing, but the increased complexity of modern electronic products has forced users to turn to automatic testing. Take, for example, the widespread use of microprocessors. These products turn out to be technologically difficult to test due to the myriad functions they can perform. The list of possible responses to the instruction set for a microprocessor can be quite long. The way in which microprocessors typically are placed in a circuit tends to inhibit fault isolation. As shown in Fig. 12.24, a topological loop is used between the microprocessor and memory elements to allow it to call up each new instruction and to store data. To complicate matters, transmission and retrieval of data are commonly done over a bidirectional bus, making demands on the ATE system to sometimes be a "talker" and at other times to be a "listener."

Numerous ATE systems have been developed to test bare PCBs, components, and fully loaded PCBs with components soldered in

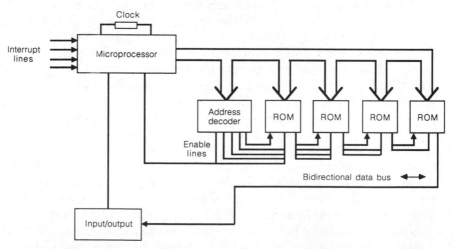

Figure 12.24 The typical architecture used in a microprocessor-based product. Testing is complicated by the topological loop between the microprocessor and its memory elements (RAMs and ROMs).[1]

place. To describe these systems, their various levels of performances, and their manufacturers would exceed the scope of this chapter; however, numerous published papers, given in the Bibliography, can be reviewed for this information. In any event, prior to purchasing any type of ATE, a return-on-investment analysis relating to the costs of subcontracting is highly recommended.[1] Today, most small companies are avoiding the high cost of purchasing and maintaining ATE and are only performing static functional testing at elevated temperatures while the product's subassembly and components are tested by subcontractors who concentrate only on testing and fault diagnosis.

12.4 Fixturing

Interfacing between the test systems and the assembly being tested is accomplished through a fixture as simple as a PCB edge connector for functional testing or as complex as a bed-of-nails fixture with a vacuum hold-down and spring-loaded probes to allow access to virtually any test land on the board. This later type of test fixture is used for in-circuit testing and can be accomplished using either a one-sided

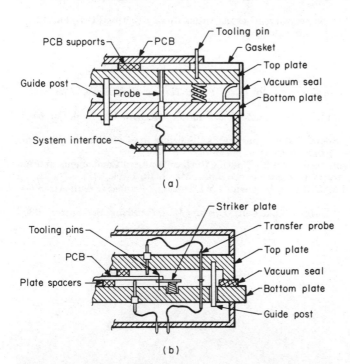

(a)

(b)

Figure 12.25 (a) The single-sided and (b) clamshell test fixtures. (*Courtesy Hewlett-Packard, Manufacturing Test Division.*)

probe fixture or a double-sided probe fixture more commonly called a clamshell fixture. Both types of fixtures are illustrated in Fig. 12.25. These test fixtures are becoming a major stumbling block in terms of cost and design turnaround time. Costs can escalate up to thousands of dollars for complex assemblies, and turnaround time can be several months between the initial board analysis and delivery. For these reasons, some industry observers claim that manufacturers will be forced to revert to functional testing, thus avoiding the need for expensive fixturing. The drawbacks of functional testers offered by proponents of in-circuit testing include extensive time and cost for programming and their relatively poor diagnostic capability.[1]

While these concerns are being tossed around, companies such as GenRad, Teradyne, and Zehntel have developed universal clamshell fixtures aimed at solving the dilemma of testing SMAs. The cost savings in using these universal fixtures can be substantial because the user requires only a single unit, regardless of the number of PCB types being tested.

References

1. Kenneth Jessen, "Overview: Approaches for Automatically Testing PCBs, Parts 1-4," *Assembly Engineering,* March 1982.

Bibliography

Bierman, Howard, "Finding a Way Out of the Board-Testing Nightmare," *Electronics,* December 9, 1985.

Coombs, Clyde F., *Printed Circuit Handbook,* 3rd edition, McGraw-Hill Book Company, New York.

Keeler, Robert, "Bare Board Testers Find the Faults before the Cost Soars," *Electronic Packaging and Production,* January 1984.

Love, Gail F., and Fennimore, John E., "Testing Surface Mounted Components and Assemblies," *IPC Technical Workshop Proceedings,* March 10, 1984.

Mangin, Charles, and McClelland, Stephen, "SMT Testing," *Printed Circuit Assembly,* January 1988.

Spada, Thomas, et al., "Tester Flunks Out," *Circuits Manufacturing,* September 1986.

Index

Note: References in *italics* are to illustrations and tables.